THUNDERSTONES

University Museum Special Publication 1

University of Arkansas Fayetteville

THUNDERSTONES

A Study of Meteorites Based on Falls and Finds in Arkansas

Derek W. G. Sears

The University of Arkansas Press
Fayetteville 1988 London

Manufactured in the United States of America
92 91 90 89 88 5 4 3 2 1

Designer: Brenda Zodrow
Typeface: Linotron 202 Electra
Typesetter: G&S Typesetters, Inc.
Printer: Thomson-Shore, Inc.
Binder: John H. Dekker & Sons, Inc.

The paper used in this publication meets the minimum requirements of the American National Standard for Permanence of Paper for Printed Library Materials Ze9.48-1984. ∞

Library of Congress Cataloging-in-Publication Data
Sears, Derek W. G., 1948–
Thunderstones: a study of meteorites based on falls and finds in Arkansas.
(University Museum special publication, University of Arkansas, Fayetteville; 1)
Bibliography: p.
Includes index.
1. Meteorites—Arkansas. I. Title. II. Series.
QB755.5.U6S43 1987 523.5′1 87-5086
ISBN 0-938626-94-9 (pbk.)

Contents

Illustrations

Preface

When I came to the University of Arkansas in 1981, my first impulse as a meteorite specialist was, naturally enough, to look up the names of meteorites that had fallen in the state of Arkansas. Their number, I discovered, is not large—just thirteen—but they include a fascinating mixture of meteorite types as well as some extremely important and well-known individual meteorites. On contacting the Curator of Geology and the Director of the University of Arkansas Museum, I was pleased to learn that the Museum would be interested in publishing a small book on the subject. After gathering the information, I wrote to several persons for help in providing particular illustrations and was overwhelmed by their cooperation. Several of them took pictures specifically for this work. The result is a much expanded and certainly a much better study than I alone could have compiled, containing many previously unpublished photographs that even specialists might find helpful. Such exuberant response on the part of those I contacted reflects, I think, the desire of many specialists today to bring their subjects to a wider readership than the technical journals can reach. Those who have provided figures or permission to reproduce published figures are: Vagn Buchwald

(Technical University of Denmark, Lyngby), Roy Clarke, Jr. (Smithsonian Institution, Washington, D.C.), Linda Hertz (*Scientific American*), Gero Kurat (Naturhistorisches Museum, Vienna), Brian Mason (Smithsonian Institution), Carleton Moore (Arizona State University), Harvey Nininger (Westminster, Colorado), and Ed Scott (University of New Mexico); I extend to them all my warmest thanks. I also have to thank the members of my own research group for making our research especially exciting: Karen Weeks, Fouad Hasan, Kyle Guimon, Brad Keck, John DeHart, Munir Haq, and Chris Hartmetz. Special thanks go to Brad and Karen, Anne Marie Candido of The University of Arkansas Press, and Ed Scott, who helped to clarify much of the text of the present work. Lastly, I thank the state of Arkansas for providing me one of the most beautiful environments on earth to pursue my studies of some of the most fascinating objects not of our earth.

I Introduction

Meteorites are pieces of stone, or sometimes pieces of metal, which have fallen to earth from space. They may be found lying on the ground, plowed up in the fields, or occasionally housed in the collections of rock hounds. Their unusual appearance and high density make them conspicuous, and many of them are magnetic.[1] Occasionally their fall from space is witnessed, and the effects are most spectacular. For ten seconds or so, the light of the fireball may turn night into day. The meteorite breaks up high in the atmosphere, and each fragment produces a sonic boom, so that a series of sonic booms is produced at the rate of machine-gun fire. Windows may be thrown open and objects may fall from shelves—effects similar to those produced by earthquakes. At one time it was thought that these curious, heat-blackened rocks were terrestrial rocks that had been struck by lightning, and they were sometimes called "thunderstones." This notion was dispelled, however, by early chemical work when it was discovered that no matter where they fell, the stones were very similar in

[1] See Appendix 2 concerning recognition of meteorites.

appearance and composition. Whether a fall or a find, meteorites are usually named after the place of fall, a custom which dates back to the eighteenth century, perhaps even earlier.

As exciting as the fall of a meteorite is, in many respects the study of meteorites is even more exciting. They are unlike any rocks found on the earth or on the moon,[2] and they are much older by far than terrestrial or lunar rocks—as old as the sun and the solar system (4,600 million years)—and, strangely enough, they resemble the sun compositionally. They also contain internal structures not seen elsewhere and which would not have survived the processes experienced by terrestrial and lunar rocks. Clearly, they are virtually unaltered relics from the earliest days of the solar system. They are our only tangible source of information about the processes which occurred then, and which led to the formation of the sun and the planets. They may also provide clues to processes which are even older than our solar system. But meteorite studies still contain many open questions, and many missing links, many puzzles remain. Those persons involved in the study of meteorites and early solar system processes, however, have an insatiable appetite for new meteorites, and many times in the past a new and different meteorite has produced an avalanche of new data and ideas.

The last British Museum Catalogue of Meteorites, one of the best sources of meteoritic statistics, lists over 2,784 in the world's museums, a figure that is now much too low.[3] The thirteen Arkansas meteorites obviously represent a small proportion of the total, but they are a richly varied group, containing some extremely important meteorites. Many of them are familiar to

[2]This is a generalization, although at least 99% accurate. There are currently four meteorites which are known to have come from the moon, and eight which very probably came from Mars.

[3]It does not include the several thousand meteorites recently discovered in Antarctica, where the movement of ice sheets and the evaporation and abrasion of ice concentrate huge numbers of meteorites in very small areas.

most specialists in the field, and are still the subject of active research.

The Arkansas meteorites are listed in Table 1 along with some data concerning their recovery, classification, and description. Four of them are stones, eight are masses of metallic iron-nickel alloy, and one is a rare example of a meteorite that is an equal mixture of stony material and metal. Six were actually observed to fall, and seven are finds. The first meteorite recovered in Arkansas was found at Joe Wright Mountain in 1884 and the biggest is the Paragould meteorite, which was seen to fall in 1930.

In this small work I will review the meteorites recovered in Arkansas, describing their recovery, their major properties, and some of the results of their study. It is an incomplete work, as their study continues. Moreover, it is a difficult task to do satisfactorily because the study of these thirteen touches on every aspect of modern meteorite studies and this is a vast research area. I have therefore attempted, where necessary, to review very briefly current meteoritic knowledge; several detailed treatments are listed at the end of the text.

II Fall Phenomena

The Paragould, Cabin Creek, Fayetteville, Norfork, Miller, and Success Meteorites.

Meteors, or shooting stars, are seen to streak silently across the sky every evening of the year, and they are fun to watch. The fall of a meteorite to earth is an altogether different affair. Meteorite masses are typically about one million times the mass of the meteor-forming object, and the effects are among the most spectacular natural phenomena witnessed by man. Far from being fun to witness, they are sometimes described as terrifying. It is no surprise that, in the past, reports of meteorite falls sounded like superstition and that for centuries the intelligentsia refused to believe them.

Six of the thirteen Arkansas meteorites were seen to fall. The fall of one of them, Paragould, is probably the best known and most studied of all meteorite falls except those which were actually photographed. Camera networks have been especially established to photograph meteor trails and, to date, three meteorites have fallen into the area occupied by the networks. The Paragould meteorite is one of the largest of any whose fall was observed. It appeared at a time when there was much interest in fall phenomena and meteorite recovery, and when much was written about them. Following its fall, two men launched systematic surveys involving the press, questionnaires, and detailed interviews: Charles C. Wylie, a Professor of Astronomy at the University of Iowa, and Harvey H. Nininger, who has done much to recover new meteorites.

The Paragould meteorite fell at 4:08 A.M. on February 17, 1930. Some observers at St. Louis and an engineer on the train between Topeka and Burlingame, Kansas, thought they saw an airplane falling in flames. After the fireball, other observers heard explosions "like a sharp peal of thunder, or a blast of dynamite." But the eyewitness reports gathered by C. C. Wylie and H. H. Nininger cannot be improved by shortening or paraphrasing: they demand verbatim reproduction. The following was taken from an account by Wylie:

> Near Beach Grove, Arkansas, two young men, Charlie Norman and Willie Allison, were going fishing. They were driving in a farm wagon when the landscape lighted up; it seemed to them brighter than day. Looking up they saw a ball of fire with a tail to it coming from the northeast. It passed overhead, and went out at an altitude of about 20 degrees in the southwest. The team, accustomed to bright automobile lights, paid little attention to the display, and the boys drove on without change of pace for perhaps a hundred yards. Here, an explosion which jarred things like an earthquake startled the men and caused the horses to plunge. The first blast seemed to come from about where the meteor had disappeared, and following this a roar as though a big train were passing rolled back along the path of the meteor. It crashed back to overhead in "no time," and then on to the northeast, the rumbling being audible for perhaps half a minute. At this point, according to Charlie, Willie wanted to turn around and go back home. However, when they had quieted the team and talked things over, the two men decided to go on with their fishing trip.[1]

In the same article, Wylie also presents one of the most poignant tales concerning a meteorite fall when he describes the Paragould descent:

[1]C. C. Wylie (1980), *Popular Astronomy* **28**, 388.

> At Poplar Bluff, Missouri, a father was returning from a trip to the doctor. His child was seriously ill, and his nerves were worn from loss of sleep and worry. In his own words, "I was walking along the road and all at once the earth lit up as light as day. I looked up at the moon and saw a ball of fire with a long tail of fire to it coming out of the east. It looked like it was going to hit me so I run to get out of its way and just before it got to me it burst into flames and went out. In about one minute the earth shook like an earthquake. When I got in the house my wife was crying and said she believed that was a sign our baby was going to die, and we lost our baby four days after that happened."[2]

The result of a synthesis of the eyewitness accounts was that a brilliant meteor—between four and six times the size of the full moon—was first seen over southern Indiana, fell to the southwest moving parallel to the Ohio River, then passed over southern Missouri and, at a height of 15 km, exploded to produce three major fireballs and innumerable sparks. The fireballs disappeared over Paragould, Arkansas, while still at a height of 8 km. The burst was seen over 120 km away. The visual phenomena were followed by detonations and a roar.

Although three fireballs were seen, only two stones were recovered. A few hours after the fall, R. E. Parkinson, a farmer near Finch, found a hole 50 cm deep with a 36.3 kg (80 lb) meteorite at the bottom. After being exhibited in the high school at Paragould, the meteorite passed into the hands of S. H. Perry and was then transferred to the U.S. National Museum in Washington, D.C. Four weeks after this find (March 16), W. H. Hodges, another farmer, discovered a 2.4 m hole on a neighbor's land a few miles southwest of Finch, where a second fragment of the same meteorite, measuring 40 × 105 cm and weighing 370 kg, was recovered (Fig. 1). Soil had been thrown 9 m, and occasionally 50 m, by the impact. The meteorite was purchased by

[2]C. C. Wylie, 387.

Nininger, who sold it to Stanley Field, who donated it to the Field Museum of Natural History, Chicago, where it now rests.

Because the fall of the Paragould meteorite was so well observed, meteorite specialists could determine the path of the fireball through the atmosphere and thereby calculate its orbit in space. There are only a handful of meteorites for which such calculations are possible, and at best they are only approximate because they depend heavily on the observations of untrained and often startled observers. Considerably easier to calculate are orbits determined from photographs but, unfortunately, there are only three such orbits (shown in Fig. 2, along with the orbit for the Paragould). The dimensions of the orbits of the Paragould and the Pribram meteorites—the latter which fell in Czechoslovakia in 1959—are very similar. They differ only in their orientation in space. All four orbits are similar to those of the Apollo and Amor asteroids, which are chunks of rock that orbit the sun in eccentric paths which take them near to, or across, the orbit of the earth. In terms of the life span of the solar system, these orbits are very brief. Typically they can exist for only about ten million years, after which the meteorites are catapulted into a different orbit and ultimately out of the solar system by the gravitational fields of the sun and planets. The meteorites, therefore, must have formed in different orbits and have been thrown into these orbits recently. (Meteorite formation locations are discussed in Chapter IX.)

An equally colorful account of the final stages of the fall of the Cabin Creek meteorite has also been documented. G. F. Kunz, a mineral dealer from New York who played an active part in recovering meteorites in the United States in the late nineteenth century, recorded details of the fall.

> The noise was heard 75 miles away and was likened to a loud report followed by a hissing sound as if hot metal had come in contact with water. It caused a general alarm among the people

A

> and teams of horses twenty five miles distant becoming frightened, broke loose, and ran away, and in Webb City, Franklin Co., on the south side of the Arkansas River, a number of bells kept on sale in a store, are said to have been caused to tinkle. Cabin Creek is on the north side of the Arkansas River.[3]

The meteorite fell 75 m from the house of Christopher C. Shandy in Cabin Creek (now Lamar), Johnson County, Arkansas.

> Mrs. Shandy states that about three o'clock on the afternoon of the 27th of March, 1886, while in her house she heard a very loud report, which caused the dishes in the closet to rattle and which she described as louder than any thunder she had ever heard. At first she thought it was caused by a bombshell, and ran out of the house in time to see the limbs fall from the tops of a tall pine tree, which, she says, stands about 75 yards from her dwelling.[4]

[3]G. F. Kunz (1887), *American Journal of Science* **33**, 495.
[4]G. F. Kunz, 495.

B

FIGURE 1. Photographs of (A) the 408 kg stone meteorite (*see facing page*) which fell in Paragould, Arkansas, February 17, 1930, and of (B) University of Iowa astronomer, C. C. Wylie, standing proudly by the 8-foot hole it created. These photographs first appeared in *Scientific American* **144**, 180–181 (1931). (By permission of the editors of *Scientific American*)

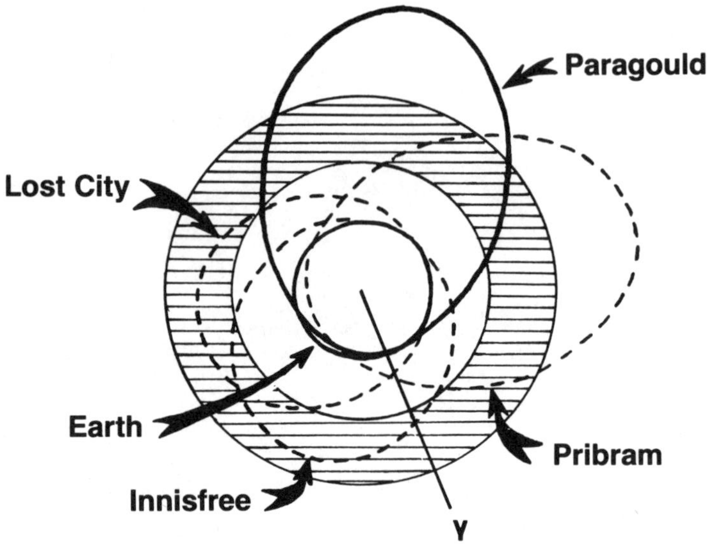

FIGURE 2. Orbit of the Paragould meteorite calculated by H. E. Nelson and W. J. Thomsen, *Popular Astronomy* 55, 448 (1947) from eyewitness observations, compared with three orbits recently determined from photographic records of meteorite falls.

At about 6:00 P.M. Mrs. Shandy, her husband, and their hired man discovered a large hole around which fresh soil had been thrown 9 m into surrounding trees. They dug, and, after "a steam or exhalation arose," found an iron meteorite "as hot as the men could well handle."

The colorful history of the Cabin Creek meteorite did not end there. Mr. Shandy sold the meteorite to B. Caraway, the mayor, who also witnessed the fall. Mr. Caraway, in turn, sold it to Colonel J. C. Betten, a lawyer from Eureka Springs. Betten published pamphlets describing the iron and offered a view of the "veritable wonder" for a fee of twenty-five cents. Kunz purchased the iron, and it passed, with the rest of his collection of ninety-one meteorites, to the Naturhistorisches Museum in Vienna around 1890.

The 42 kg Cabin Creek meteorite is one of the most photographed, yet least studied, meteorites (Fig. 3). The two facts are related; it would take a courageous curator to cut the beautiful iron, though it might well prove scientifically valuable to do so. The Russellville *Democrat* newspaper (April 29th, 1886) reported the fall in breathless, yet understandable, prose:

> We looked at the strange thing and wondered what it was and where it came from. The noise it made when it struck the earth's atmosphere on the 27th of March and came whizzing to earth near Knoxville, will never be forgotten, neither will anyone who looked at it ever forget it.

It is a spectacular plate-shape, 44 × 39 cm, and only 7.5 cm thick. One side is flat with large shallow pits; the other is deeply ridged and dented with 3–4 cm diameter depressions sometimes called "regmaglypts," which are thought to have been formed by turbulent sculpturing by air currents during atmospheric passage. Its overall shape reflects the orientation of the meteorite during fall. Meteorites adopt a maximum drag configuration during fall, so that a plate-shaped object presents its largest face to the direction of flight. There are many fine hairs and other textures resulting from the flow of liquid metal which became "frozen" at the moment of impact with the dense, lower levels of the atmosphere.

Miller is another Arkansas meteorite which remained intact during fall and became an oriented stone (Fig. 4). It has been described as heart-shaped, about 30 cm across and 15 cm thick (weight 16.5 kg), and, except for a few small places where chipping has occurred, is uniformly covered with a thin layer of black crust. It is the normal shape for an oriented stone meteorite, and many similar examples are known. It is no accident that the shape of the meteorite is reminiscent of spacecraft heat shields. The low thermal conductivity, high heat content, and the man-

FIGURE 3. The iron meteorite which fell at Cabin Creek, now Lamar, Arkansas on March 27, 1886. The mass is 44 × 39 × 7.5 cm and weighs 48.2 kg. The face shown was at the front during atmospheric passage and has been excavated by atmospheric sculpturing to produce numerous depressions. The rear face is relatively smooth. (By permission of G. Kurat, Naturhistorisches Museum, Vienna)

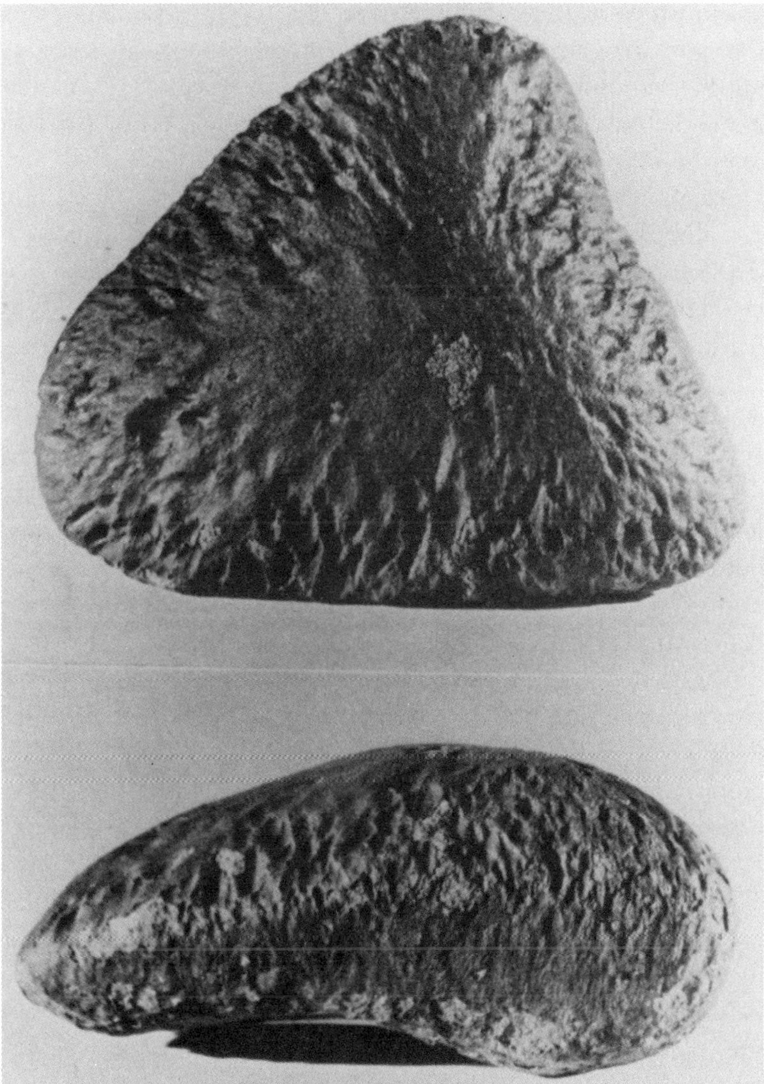

FIGURE 4. Photographs of the Miller, Arkansas, meteorite from two sides. The smoother, larger face traveled foremost during atmospheric passage. (By permission of B. Mason, Smithsonian Institution)

ner in which heat is rapidly removed by run-off of molten material cause the interior of the stone to remain remarkably cool during the atmospheric passage. Depths as little as 0.5 cm in the meteorite have never been heated above 250°C despite the fact that the amount of heat generated is enough to dissociate and ionize molecules of air. The black crust, or fusion crust, typically a millimeter or so in thickness, is a mixture of glass, magnetite (Fe_3O_4), and the occasional high-melting-point mineral. The surface texture of the crust at various sides of the meteorite differs considerably, reflecting the motion of the fluid material during atmospheric passage: smooth at front; warty or rough at the rear; with long, fine, drawn-out hairs of material at the sides.

The fall of the Miller meteorite seems to have been a relatively quiet affair, according to the few words offered by C. A. Reed.[5] The meteor was likened to the flight of a white pigeon, but moving much faster. It was reported to have raised a cloud of dust in a dirt road and made a 50 cm hole. It is now in the American Museum of Natural History, New York.

The meteorite most recently witnessed in Arkansas fell on December 26, 1934, at 11:58 A.M. just 6 km west of Fayetteville. Because of its time of descent, visual effects do not feature as prominently in the accounts as they do for the Paragould. D. P. Richardson recounted that while cutting wood, Pearl Conley and Ray Brisco heard an explosion, then a rumbling and a dull thud.[6] The air around them was disturbed "as if the breeze just died" and cattle in a nearby field behaved strangely. Upon investigation they found a 30–38 cm hole from which they excavated a warm, 2.3 kg meteorite at 50 cm depth. At greater distances from the fall site, observers heard several sharp explosions and lesser explosions "which faded into rumbling," and there were several reports by observers who thought they saw falling airplanes. Six days later

[5] C. A. Reed (1931), *Natural History* **31**, 109.
[6] D. P. Richardson (1935), *Popular Astronomy* **43**, 384–385.

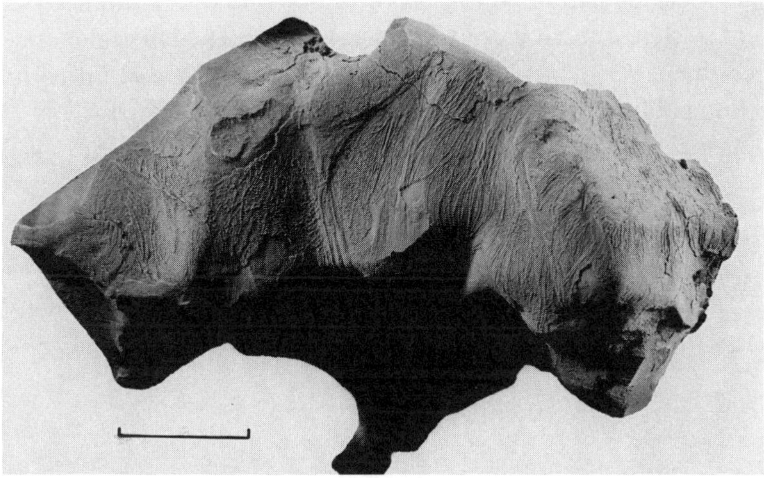

FIGURE 5. The iron meteorite from Norfork, Arkansas. Patterns caused by the flow of liquid metal as the meteorite passed through the atmosphere are readily apparent. Scale bar 2 cm. (By permission of R. Clarke, Jr., Smithsonian Institution)

a second 108 g stone was found 6 km to the northeast of the fall site of the larger mass. The larger mass is entirely covered in fusion crust and shows signs of being oriented in the atmosphere. The smaller stone shows more than one generation of crust, and one face, presumably exposed later in the flight than the others, shows a less developed crust. Both stones were collected from their finders by Davis Richardson, Chairman of the Mathematics Department at the University of Arkansas in Fayetteville, and are now part of the University of Arkansas Museum collections.

Virtually nothing is known about the fall of the two remaining Arkansas meteorites: Success and Norfork. The fall of the Norfork meteorite was observed by Isaac C. Luther, who the next day found a 1.05 kg iron in a hole 120 cm deep. The meteorite has beautiful stria (small strings of material) caused by the movement of molten material on its surface during fall (Fig. 5). The mass became part of Nininger's collection and is now at the Center for Meteorite Studies in Tempe, Arizona. The fall of the Success meteorite was announced briefly in the *Meteoritical Bulletin*, published by the Soviet Academy of Sciences, for 1935. The 3.5 kg mass is now housed in the Smithsonian Insitution.

III The Finds

The Joe Wright Mountain, Sandtown, Hatfield, Calico Rock, Newport, Western Arkansas, and Slovak Meteorites.

Recovering meteorite finds in Arkansas is not easy, even though the United States as a whole is a particularly rich source of them, probably because of its geographic and demographic history. Meteorites have fallen on North America since its creation, and until the nineteenth century the continent was sparsely populated. They therefore had much time to accumulate on the vast, dry, open prairies. By the end of that century these plains were relatively well populated and well cultivated, and recovery of meteorites was almost commonplace. Arkansas is divided by a line northeast to southwest above which are mountains and below which is the Mississippi delta plain (Fig. 6). Neither area is well suited to meteorite recovery, but the river plains are less so because the waterlogged nature of the terrain engulfs and destroys meteorites. Only one meteorite, the recently recovered Slovak, has been recovered from this area. The mountains provide a somewhat more hospitable environment, but agriculture is difficult and the land is virtually undisturbed. This region has not suffered the scrutiny of the prairie lands, which are annually subjected to plowing, sowing, tending, and harvesting.

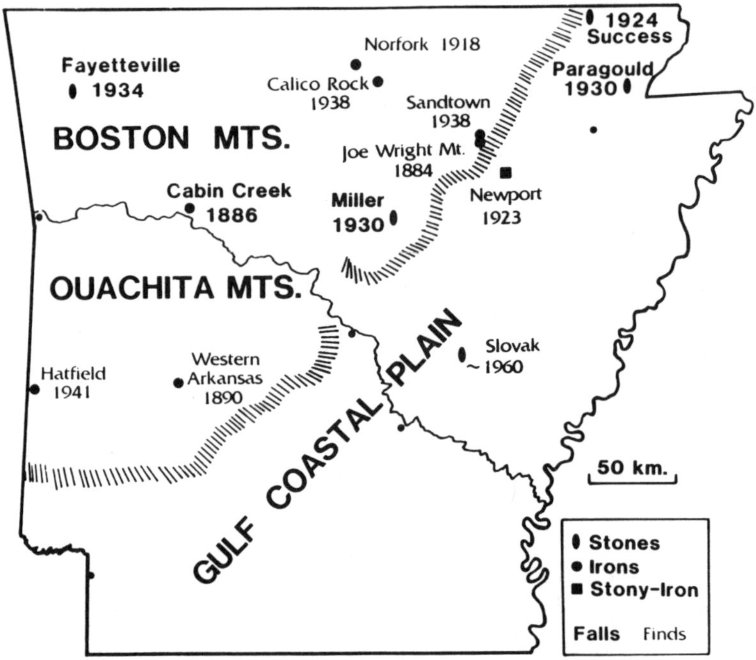

FIGURE 6. Map of Arkansas showing the find or fall sites of the thirteen Arkansas meteorites.

Probably the best known of the Arkansas meteorite finds is Joe Wright Mountain. The conditions of the find of this meteorite were given in 1886 by W. E. Hidden who reproduced a letter from Mr. John Hindman of Elmo, Arkansas. Hindman wrote:

> This meteor was found about the last of June, 1884, by my stepson George W. Price, on a mountain known as the "Joe Wright Mountain," a small eminence situated about seven miles east of Batesville (Independence County, Arkansas). The soil there was cut into deep gullies, which, farther down the mountain side converged into one. It was where these gullies met that the meteor was

> found. The town of Sulphur Rock is about three miles distant, southwest, from the place of the discovery.[1]

According to Buchwald and Hidden, the meteorite is a 48.2 kg iron measuring 40 × 20 × 15 cm (Fig. 7).[2] Its surface is too weathered to show any atmospheric sculpturing except perhaps 2–5 cm depressions which may be remnants of remaglypts. The history of the iron is largely undocumented, but the meteorite was in the Vienna collection by 1889. It is now well studied and well distributed; at least eighteen museums world-wide have samples of the iron.

A few years after the discovery of the Joe Wright Mountain meteorite, a second mass of iron meteorite, weighing 9.4 kg, was found in the same area (in fact, 16 km north and 10° west). Within a year of discovery, it became part of the collection of the Field Museum of Natural History in Chicago. From the time of its discovery until recently, this meteorite was thought to be a fragment of the Joe Wright Mountain meteorite, but Buchwald has found that its internal structure is very different from that of the older iron and is altogether a separate meteorite, which he calls Sandtown.[3]

The confusion over the Sandtown and Joe Wright Mountain meteorites is an example of a perpetual problem facing meteorite specialists, and particularly curators, who are brought several pieces of meteorite from the same location. Are they fragments of a single meteorite or of different meteorites? The problem is particularly acute for the recent discoveries of meteorites in the Antarctic. Over 4,000 individual pieces of meteorite have

[1] W. E. Hidden (1886), *American Journal of Science* **31**, 461.
[2] V. F. Buchwald (1975), *Handbook of Iron Meteorites*, 697. W. E. Hidden (1886), *American Journal of Science* **31**, 460.
[3] Buchwald, 697.

FIGURE 7. Engraving of the 42.7 kg iron meteorite found on Joe Wright Mountain in 1884 as it appeared in W. E. Hidden's original description of the iron in *American Journal of Science* **31**, 461 (1886). The ribbon indicates the location of a hole probably made when an inclusion melted out of the meteorite during atmospheric passage.

been recovered there, but many look so alike that they must be part of a single meteorite which fragmented in the atmosphere. Although oriented single falls, like those of the Cabin Creek or Miller meteorites, are not uncommon, many meteorites fragment low in the atmosphere and scatter fragments over an ellipse whose major axis lies in the direction of travel. Because of their greater momentum, the heaviest fragments travel farther than lighter fragments before atmospheric resistance slows them down. There is therefore a steady increase in mass along the ellipse in the direction of travel. The L'Aigle meteorite, which fell in France in 1803, produced 3,000 fragments and the Pultusk, Poland meteorite produced over 100,000 separate stones. The problem becomes more than one of curatorial interest when the less common classes are involved. For example, nearly half the so-called

FIGURE 8. Photograph of two cut and polished faces of the Slovak meteorite showing the large metal nodules present in this meteorite. It is widely assumed that an impact in space caused the small metal grains to coalesce into large nodules.

type 3 ordinary chondrites known may be fragments of a single meteorite known as Allan Hills A77011, and this fact must be taken into account when studying this type of meteorite.

The discoveries of the Calico Rock and Hatfield meteorites seem to have prompted little excitement, and few details have been recorded. A certain A. Harmon found a peculiar brick-shaped 7.28 kg mass of iron near Calico Rock, Izard County, in 1938 (Fig. 8). It was acquired in 1964 by R. A. Orti and unfortunately remains in private possession, essentially unavailable for serious study, though small samples have been sent to the national museums in Washington, London, and Copenhagen. The curious shape of the mass is the result of breakup along crystal faces (see Chapter VI). The Hatfield meteorite first appeared in the literature as an entry in the catalog of Harvey H. Nininger's collection.[4] A 21 g specimen was said to have been found in 1941. Along with half of Nininger's collection, it passed to the Center for Meteorite Studies at Arizona State University in 1970.

The history of the Western Arkansas meteorite is only marginally better documented than that of the previous two meteorites. G. M. Merrill reported that by 1926 it was a part of the F. A. Canfield collection of meteorites bequeathed to the Smithsonian Institution, where it resides today.[5] Canfield's label, attached to the specimen, read: "Mr. Wilson said a native mountaineer brought the specimen to him in Joplin, Montgomery County, and that it was . . . presented to I. Price Wetterill, June, 1890." V. F. Buchwald describes the meteorite as an 8 × 6 × 6 cm mass weighing 1.75 kg.[6] It is apparently part of a larger mass, now lost, and bears numerous hammer and chisel marks. Its internal structure indicates that the entire mass has been artificially heated to ~800°C (see Chapter VII).

The only stone meteorite found in Arkansas is Slovak. Frank Halligan, a farmer who lives near Carlisle, found a large conspicuous stone lying in a fence row on prairie-type rolling ground near the small town of Slovak in 1960 or 1961. For over twenty years he kept the meteorite in his mechanical workshop, but to his knowledge it did not suffer any reheating or mechanical treatment. In 1982 Jim Wescott, a meteorite dealer from Sedona,

[4] H. H. Nininger and A. D. Nininger (1950), The Nininger Collection of Meteorites, Winslow, Ariz.
[5] G. M. Merrill (1927), Proceedings of U.S. National Museum, 72, art 4., 1–4.
[6] V. F. Buchwald(1975), *Handbook of Iron Meteorites*, 1304–1305.

FIGURE 9. Photographs of the stone meteorite found near Slovak, Arkansas, in the early 1960s. It is the only stone meteorite to be recovered in Arkansas which was not actually seen to fall. The stone measures 23 × 23 × 12 cm.

Arizona, recognized the stone as a meteorite, and in November 1983 the author persuaded Mr. Halligan to part with it. The main mass was donated to the University of Arkansas Museum. The find coordinates are 34°38′7″ N, 91°31′39″ W.[7]

Before being cut, the stone weighed 8.22 kg and was an irregular mass measuring approximately 23 × 23 × 12 cm (Fig. 9). The stone is uniformly weathered brown, but in places hints of fusion crust are present suggesting that the meteorite has undergone very little weathering. One face, however, is somewhat smoother than the other, especially at certain locations around the edges, a fact that may reflect atmospheric sculpturing. Pieces chipped from the edges of the mass reveal a black interior.

[7] Estimated from United States Geological Survey (USGS) topographical map of Parker Corner, Arkansas.

IV The Stony Meteorites

The Paragould, Miller, Fayetteville, Success, and Slovak Meteorites

The five stony meteorites from Arkansas are all members of the so-called ordinary chondrite group of meteorites. Of all the stony meteorites seen to fall, 85% are chondrites; the remainder are termed "achondrites." The difference between the two types of meteorites is primarily compositional, although the names refer to the fact that, in general, chondrites contain "chondrules," whereas achondrites do not. Chondrules are crystalline aggregates of material which are often spherical and look like once-melted droplets that have solidified. Except for most volatile elements, the chondrites resemble the sun in composition, while the achondrites are very different, since they are very poor in siderophile elements[1] and either enriched or depleted in refractory lithophile elements (calcium, aluminum, and others) compared with the chondrites (Fig. 10). There are no Arkansas achondrites, but for the sake of completeness, the achondrite classes are

[1]This classification of elements was introduced by V. M. Goldschmit in 1930 and has proved most helpful. "Siderophile" elements are those which tend to go into the metal whenever possible (e.g., nickel, cobalt, gold, and iridium); "lithophile" elements are those which tend to form oxides (magnesium, silicon, calcium, and aluminum); "chalcophiles" are elements which tend to form sulfide compounds (copper, zinc, and selenium). Atmophiles are the inert gases and nitrogen.

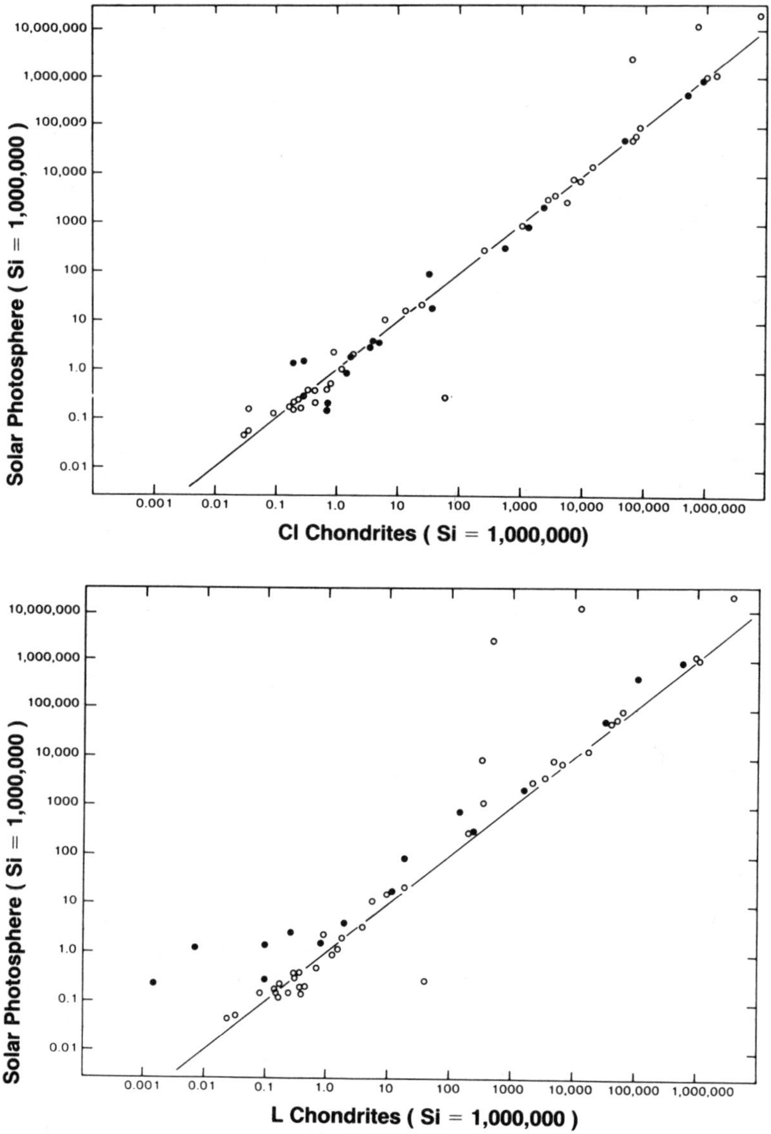

FIGURE 10. Comparison of the composition of the sun with that of several classes of stony meteorite. Compositions are on a scale with silicon equal to 1,000,000, so that an element present in 1/100th amount of silicon has a value of 10,000, and so on. The diagonal line represents agreement between the two sources. Siderophile and chalcophile elements are plotted as filled symbols, and lithophiles as open symbols. CI and L chondrites have basically the same

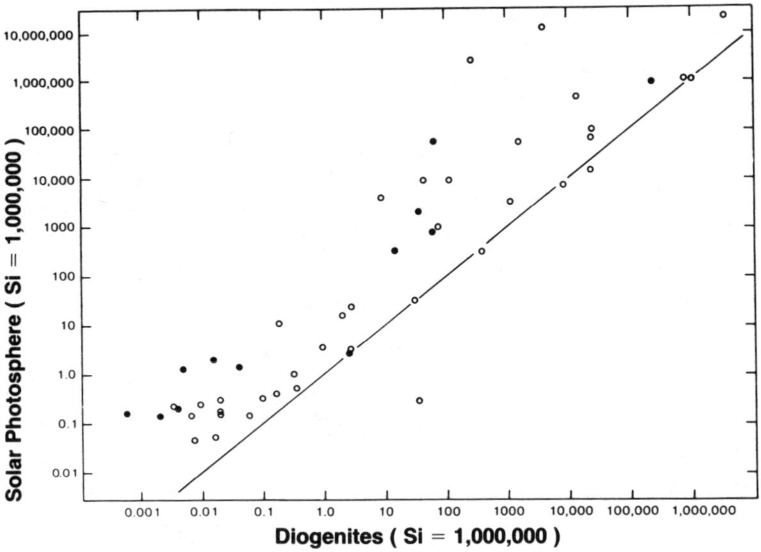

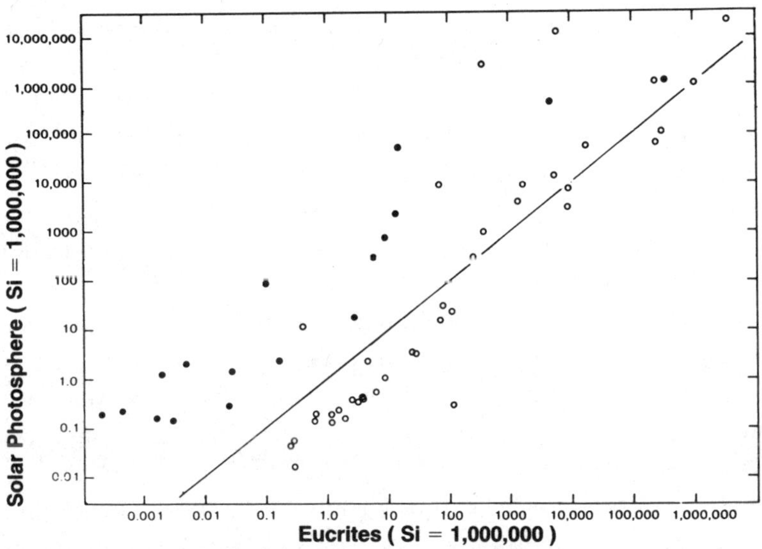

composition as the sun, especially the former, whereas the eucrites and diogenites are non-solar but in very different ways. All but a few elements (notably magnesium and silicon) are low in diogenites, whereas eucrites are low in chalcophiles and siderophiles, but high in lithophiles. (Prepared for this work from data in B. Mason, Geological Survey Professional Paper, 440–B–1, 1979, and J. E. Ross and L. H. Aller, *Science* **31**, 1223, 1976)

included in Table 2, which lists the currently recognized meteorite classes.

Despite their essentially solar composition, there are subtle compositional differences which have led to the definition of nine separate chondrite classes. The existence of such classes, and the chemical trends that produced them, offer unique clues to their formation conditions and to the processes which occurred in the early solar system.

The relevant chemical trends are three-fold. First is the amount of oxygen in the meteorite, a trend difficult to measure directly but which readily manifests itself in the mineral properties of the chondrite. At the oxygen-rich extreme are the carbonaceous chondrites with nearly all the iron combined with oxygen as magnetite or silicates, and some even contain up to ~20% water. As their name suggests, many contain carbon. The ordinary chondrites have perhaps half of their iron in oxide form and the remainder as troilite (iron sulfide, FeS) or metal. At the oxygen-poor extreme are the enstatite chondrites with all their iron in the metal and sulfide form; they are named after the abundant silicate present, which is enstatite. The second chemical trend is the bulk iron content. (This is better expressed as the iron to silicon ratio, since the ratio, unlike concentrations, is insensitive to variations in quantities other than iron and silicon). The iron-silicon ratio leads to the subdivision of ordinary chondrites into H, L, and LL chondrites (high, low, and very low iron) and the enstatite chondrites into EH and EL chondrites. The third chemical trend involves the lithophile elements. The ratio of magnesium to silicon separates enstatite, ordinary, and carbonaceous chondrites, and, when the ratio calcium to silicon is also taken into account, leads the definition of four carbonaceous chondrite classes: CI, CO, CV, and CM.[2] Hence the nine

[2]The names are taken from the initials of well-known meteorites from each class: Ivuna, Ornans, Vigarano, and Mighei.

chondrite classes—each with a characteristic bulk composition. It is worth stressing that although well-determined major elements are used to define classes, they are not all unique. Iron could be replaced by any siderophile element, magnesium with any lithophile element, and calcium or aluminum with any refractory lithophile element. Figures 11 and 12 attempt to summarize succinctly these chemical classes.

Bulk compositional data for the Arkansas stone meteorites appear in Table 3; these data may be compared with the classes as defined in Figures 11 and 12. The Paragould analysis was made by Roy using a somewhat unconventional approach, but results clearly indicate an LL chondrite classification for the meteorite. H. B. Wiik, who has probably performed more wet-chemical analyses of chondrites than anyone else, analyzed the Miller in 1964.[3] On the basis of its bulk composition, this meteorite is an H chondrite. Determinations of the bulk compositions of Fayetteville and Success do not appear to have been made, although Fayetteville has been partially analyzed in other connections. Our own analysis of Slovak was made using a technique known as neutron activation analysis. Samples are placed in the core of a nuclear reactor where the neutron flux converts certain elements into radioactive forms. The composition of the meteorite can then be calculated from the amount and energy of gamma radioactivities induced. Our data indicate that Slovak is an H chondrite.

Since the bulk compositions, especially the oxygen content, determine mineralogy, mineral abundances and composition can also be used to identify classes. The calculated mineral abundances (the norms) for Miller appear in Table 4. The major silicates in ordinary chondrites are olivine, pyroxene, and feldspar. None of these are pure compounds but solid solutions. The components of olivine and pyroxene are:

[3]H. B. Wiik (1964), *Commentiones Physica: Mathematica* **34**, 135.

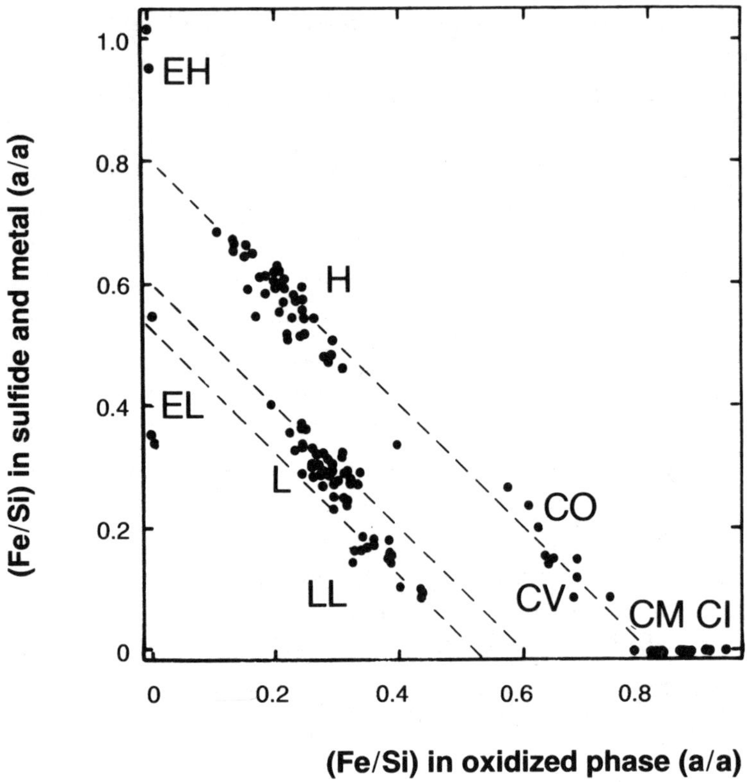

FIGURE 11. The Urey-Craig diagram in which the ratio of reduced iron (iron as the metal and sulfide, FeS) is plotted against oxidized iron (iron as silicates). Each diagonal represents a given bulk composition (i.e., a given iron to silicon ratio). The addition or removal of oxygen moves a sample down or up a diagonal, respectively. This plot resolves several classes of chondritic meteorites (basically solar in composition). (After R. W. Bild and J. T. Wasson, *Science* **197**, 58–62, 1977)

olivine: Mg_2SiO_4 (forsterite, Fo) and Fe_2SiO_4 (fayalite, Fa)
pyroxene: $MgSiO_3$ (enstatite, En) and $FeSiO_3$ (ferrosilite, Fs)

The more oxygen in the meteorite, the more iron will be in the silicate as fayalite or ferrosilite, and the less as metal. The amount of metal, therefore, decreases along the series H, L, and LL, not

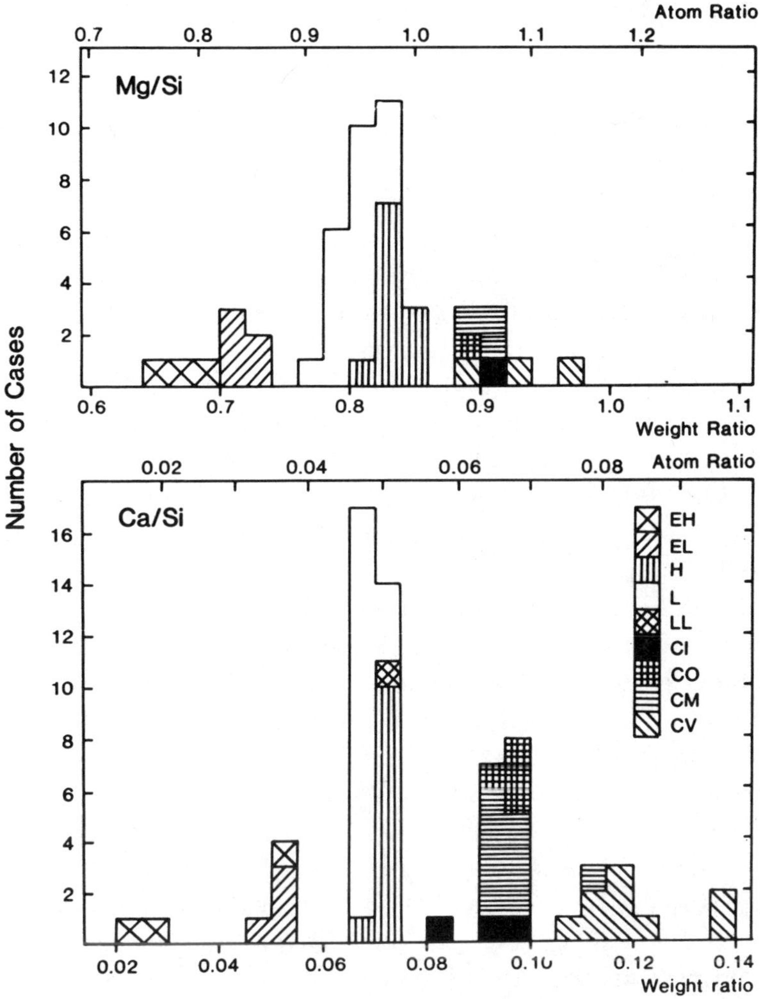

FIGURE 12. Histograms of the ratios of lithophile elements (i.e., those tending to become silicates rather than metals or sulfides) in chondritic meteorites. These ratios also resolve several classes of meteorite. (Prepared for this work from the sources given in the caption to Fig. 11; and after H. Von Michaelis, L. H. Ahrens and J. P. Willis, *Earth and Planetary Science Letters* **5**, 387–394, 1969)

just because the bulk iron decreases, but also because the amount of oxygen increases. Thus the silicate compositions define discrete groups, just as bulk compositions do (Fig. 13). The composition of the olivine in the Arkansas stones was measured by B. Mason in 1963 using X-ray diffraction.[4] He reported the following fayalite contents (in mole percent) for the Fayetteville, Miller, Paragould, and Success meteorites respectively: 19, 19, 25 and 24. These fayalite contents suggest classifications of H, H, L, and L, respectively, for these four meteorites. The mineral compositions for the Paragould meteorite were made by K. Fredriksson and co-workers using electron microprobe analysis, a technique which has largely superseded X-ray diffraction for this purpose. An electron beam is finely focused onto a polished surface of the sample, and X-rays are generated. The energy of the x-rays is determined by the kind of element present, and their intensity by the amount of that element. Fredriksson and co-workers found the mean fayalite and ferrosilite content of Paragould to be 27.6 and 22.9 mole percent respectively—normal values for an LL chondrite and in agreement with Roy's bulk analysis.[5] The minerals in the Slovak meteorite were recently measured by E. R. D. Scott, with the same technique. The mean fayalite and ferrosilite value are 17.5 and 15.5 mole percent, respectively—normal values for an H group chondrite.[6]

Besides bulk chemistry and mineralogy, detailed nuclear properties of meteorites also lead to the classification of meteorites into these same classes. There are three isotopes of oxygen with mass numbers 16, 17, and 18. To a very good approximation, the ratios of these three isotopes of oxygen are remarkably

[4] B. Mason (1963), *Geochimica Cosmochimica Acta* 27, 1011–1024.
[5] K. Fredriksson, J. Nelen, and B. J. Fredriksson (1968), in *Origin and Distribution of the Elements* (L. H. Ahrens, ed.), 457–465.
[6] E. R. D. Scott, personal communication, June 1985.

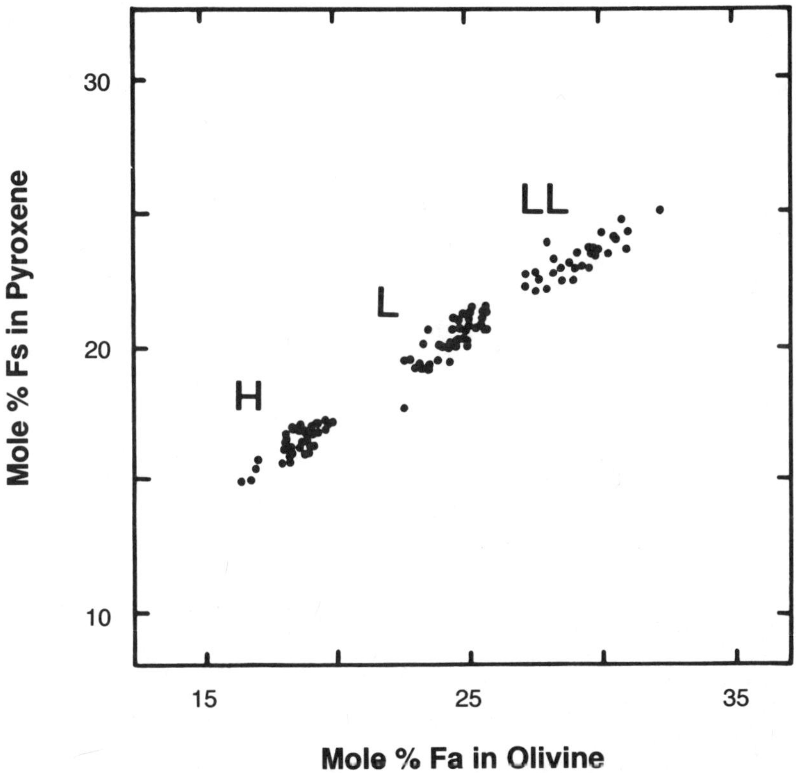

FIGURE 13. Plot of the compositions of two major silicates in chondrites, which clearly resolve the H, L, and LL chondrites. Olivine is a solid solution of Fe_2SiO_4 (Fa) and Mg_2SiO_4, while pyroxene is a solid solution of $FeSiO_3$ (Fs) and $MgSiO_3$. (Prepared for this work from data in K. Keil and K. Fredriksson, *Journal of Geophysical Research* 69, 3487–3514, 1964; and K. Fredriksson, J. Nelen, and B. J. Fredriksson in *Origin and Distribution of the Elements*, L. H. Ahrens, ed., 1968)

constant and equal to standard mean ocean water (SMOW). However, high resolution mass spectrometry reveals small differences which may be expressed as δ (parts per thousand difference from SMOW). When δ for the 0–17 to 0–16 ratio is plotted against δ for the 0–18 to 0–16 ratio, terrestrial and lunar samples

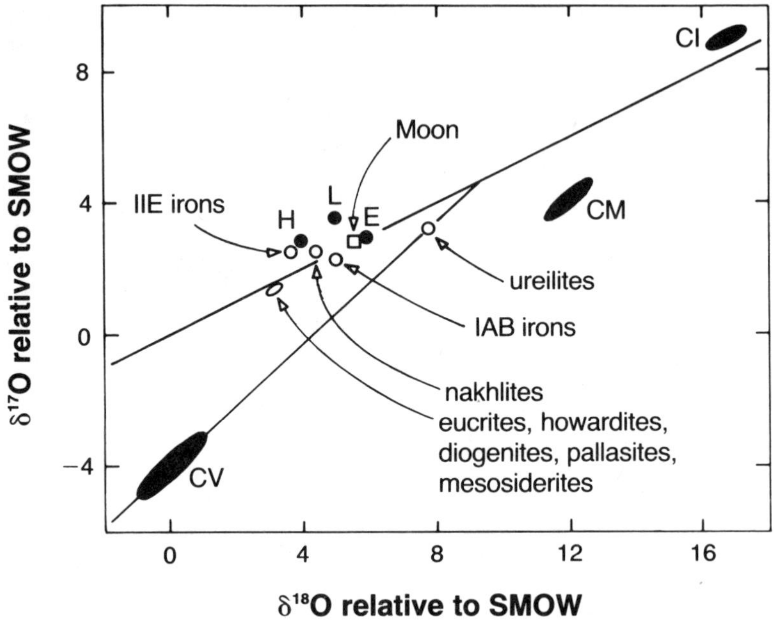

FIGURE 14. A plot which compares the ratio of oxygen-17 to oxygen-16 with the ratio of oxygen-18 to oxygen-16. (The units are δ per mil, where $\delta^{17}0 = \{(^{17}0/^{16}0)$ sample divided by $(^{17}O/^{16}0)$ standard $- 1\}1000$, and the standard is standard mean ocean water, SMOW). The slope 0.5 line corresponds to terrestrial and lunar samples which have experienced a mass-dependent process (melting, vaporization, and so on). The slope 1.0 line corresponds to inclusions from carbonaceous meteorites. The line has been explained in terms of the addition of varying amounts of almost pure 0–16 which moves points off the terrestrial line and towards the bottom left-hand corner of the plot. The meteorite classes occupy discrete fields, so this plot may be used for classification. (After R. N. Clayton, *Phil. Trans. Roy. Soc. Lond.*, A 303, 339–349, 1981)

lie on a line with slope 0.5, whereas meteorites lie in localized fields off this line (Fig. 14). Their oxygen isotope ratios can therefore be used to classify meteorites in a way somewhat analogous to bulk compositions and mineralogy. This plot is further discussed in Chapter VIII.

Bulk composition and oxygen isotope properties discussed thus far, which have led to the definition of classes, are commonly thought to have been determined by mechanical and chemical processes which occurred when the material was diffused gas and dust in the early solar system. These processes will be discussed briefly in Chapter VIII. A distinctly different kind of process, which probably occurred after the dust had gathered together and formed itself into km-sized objects (although this is disputed by some specialists), is termed "metamorphism." Technically defined, this is the solid-state recrystallization at elevated temperatures and pressures. A very wide range of metamorphic alteration has been recognized in meteorites since the last century, but a most useful overall synthesis of the effects was provided by W. R. Van Schmus and J. A. Wood. Their observations, as they apply to ordinary chondrites and are updated according to recent work, are summarized in Table 5. The definition of petrologic types for the carbonaceous and enstatite chondrites follows very similar lines.

Van Schmus and Wood suggested the definition of six "types" and that the type number be added to the class for an almost complete description of the meteorite.[7] As the type and degree of metamorphic alteration increase, the olivine and pyroxene (and also the metal) become more homogeneous in composition (Fig. 15), and the pyroxene changes its crystal form from the low-temperature to the high-temperature form (from clinopyroxene to orthopyroxene). The amount of crystalline feldspar increases and the amount of glass decreases; most likely the elements which would go into feldspar originally formed a glass which devitrified to form crystalline feldspar during metamorphism. The texture of the meteorite, as observed under the microscope, also changed;

[7]W. R. Van Schmus and J. A. Wood (1967), *Geochimica Cosmochimica Acta* **31**, 747–766.

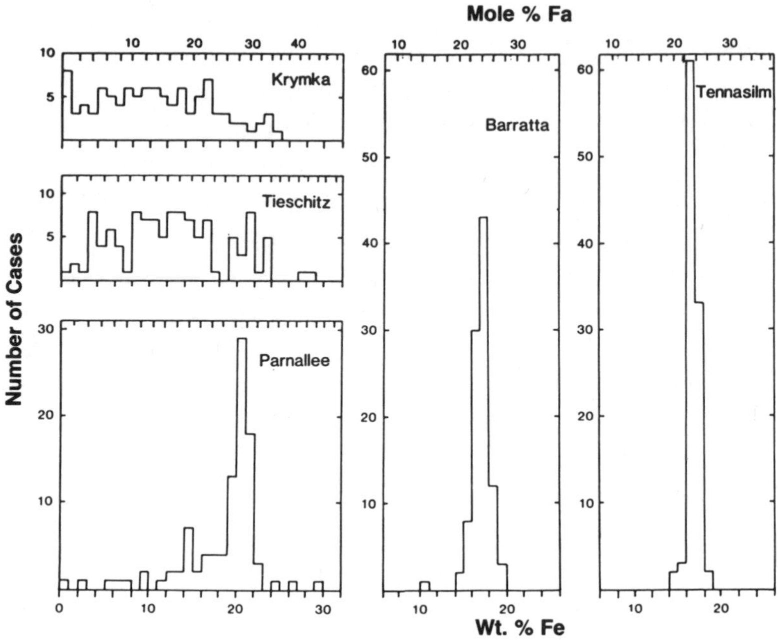

FIGURE 15. Histograms of the olivine compositions of individual grains from a variety of meteorites. Some meteorites, like Krymka, are heterogeneous and are thought to be little-metamorphosed, while others are compositionally homogeneous and highly metamorphosed—for example, Tennasilm. According to the petrologic types described in Table 5, Krymka is type 3.0, Tieschitz 3.3, Parnallee 3.7, Baratta 4, and Tennasilm 5. Type 6 have virtually no spread in the iron or fayalite content. (Adapted with permission of R. T. Dodd, from R. T. Dodd, V. R. Van Schmus, and D. M. Koffman, *Goechimica Cosmochimica Acta* **31**, 921–951, 1967)

the fine-grained dust, which forms a matrix surrounding the metal, sulfide, and chondrules, became transparent and grew into coarse crystals. At the same time, the outlines of the chondrules became "washed-out," and the chondrules began to merge with surrounding material. In recent years additional work has revealed further metamorphism-sensitive properties, most nota-

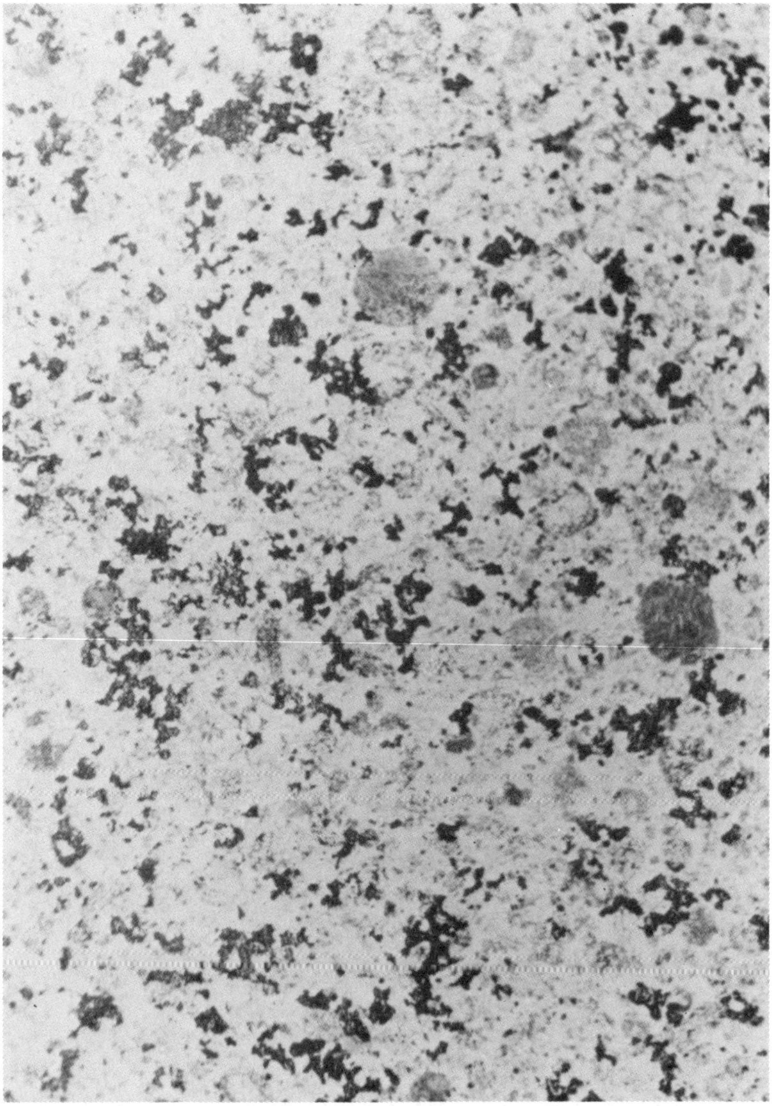

FIGURE 16. Photomicrograph of a thin section of the Miller chondrite. Gray is pyroxene, white is olivine, black is metal and troilite (FeS). Average chondrule diameter is 0.5 mm. (By permission of B. Mason, Smithsonian Institution)

bly thermoluminescence,[8] and Van Schmus and Wood's type 3 has been subdivided.

Van Schmus and Wood assigned both the Miller and the Paragould meteorites to their petrologic type 5 since they have homogeneous mineral compositions, recrystallized matrix, and the presence of coarse grains of feldspar. However, the chondrule outlines remain distinct, as evident in Mason and Wiik's photomicrograph of Miller (Fig. 16). Thus, Miller is H5, while Paragould is an LL5 chondrite. There are no petrologic descriptions of Success in the literature, but the catalog of the meteorite collection for the Smithsonian Institution lists it as an L6 meteorite. Slovak is also a type 5 chondrite on the basis of silicate heterogeneity.[9] The situation for the Fayetteville meteorite warrants further discussion, which is reserved for the next chapter.

[8]Literally, the light emitted as the sample is heated. The level of thermoluminescence, and the way it is emitted, depends on the crystal structure of the mineral feldspar and the impurities it contains.

[9]E. R. D. Scott, personal communication, June 1985.

V The Regolith Breccia

The Fayetteville Meteorite

Even the non-specialist would recognize the Fayetteville meteorite as distinct from most ordinary chondrites. On a flat, ground surface one sees a quite definite structure (Fig. 17). Instead of the usual gray, fine-grained mass interspersed with flecks of shiny metal and sulfide, Fayetteville consists of irregular fragments of normal chondrite material ranging in size from several centimeters to a dimension barely perceptible, surrounded by a darker material. The darker material is not as uniform in appearance as the lighter material, or clasts, but contains several black inclusions and even a nodule of metal a centimeter or so across. The Fayetteville meteorite is one of the best examples of a so-called regolith breccia, a rock that was once part of the very surface of a body in space.[1] The surface of this body was continually stirred up by repeated impact by micrometeorites so that fragments of normal meteorite material became imbedded in a "soil" which later became compacted to produce the dark material of

[1]The word "regolith" means loosely consolidated material on the surface of a body; the term "breccia" means a rock which is composed of a variety of smaller rock fragments.

FIGURE 17. Photographs of the stony meteorite that fell in Fayetteville, Arkansas, on December 26, 1934. The meteorite is a relatively rare example of a so-called regolith breccia which, when cut open, is seen to consist of light "clasts" of normal meteorite material in a "matrix" of darker material. The dark matrix contains a large metal nodule and numerous small black inclusions. The meteorite is 10.5 × 10.5 × 8 cm.

the final meteorite. A plethora of interesting effects have been found in this dark component.

All meteorites contain, as part of their normal chemical makeup, very small quantities of the inert gases helium, neon, argon, xenon, and krypton. When the Fayetteville meteorite was analyzed, it was found that the dark matrix contained these gases in overwhelming abundance (Table 6). Fayetteville was not the first gas-rich meteorite to be found, but it is certainly one of the most gas-rich, and has subsequently become one of the best-known regolith breccias.

All the forty or so regolith breccias currently known contain dark matrices which are gas-rich. In contrast, the light clasts have much the same low gas abundances as other, non-gas-rich meteorites. The gas is predominantly helium-4, neon-20, and argon-36—isotopes which are abundant in the sun's atmosphere. The source of the excess gas seems well established. Etching experiments have shown that the gas is trapped on the very surface of the grains and appears to be implanted solar wind. Other solar effects have subsequently been discovered. The sun emits corpuscular radiation all the time, but emission is especially associated with solar flares. The radiation has much lower energy than do normal cosmic rays and does not penetrate as far into the meteorite grains. As it passes through the grains, it rapidly deposits its energy and produces short, broad tracks of radiation-damaged material. The grains can then be etched with strong acids so that the tracks may be enlarged and made visible under the microscope. The regolith breccias contain many grains in the dark matrix which have solar flare tracks around the edges; some have tracks on one side but not the other. The tracks are unambiguous evidence that the grains were at some time directly exposed to the solar wind and that these meteorites were once part of a regolith. Because the Fayetteville meteorite is especially rich in gases, its track density is especially high ($\sim 10^{11} cm^{-2}$; other regolith breccias typically contain $\sim 10^9$ cm^{-2}), indicating that Fayetteville has one of the longest known periods of regolith activity. But some authors have argued that the gas content reflects gas loss during process(es) which transformed loose dust into solid rock (lithification), in which case Fayetteville is the least well-lithified regolith breccia.

Other effects resulting from this environment have been documented. Regolith breccias have been found with grains bearing micro-craters on their surface, produced by the impact of micron-sized micro-meteorites. The dark matrix also occasionally contains masses of glass called "agglutinates," which are associated with regolith activity. In many respects the regolith breccia meteorites are analogs of many moon rocks recovered by the *Apollo* and *Luna* missions. These moon rocks, many of which are also regolith breccias, came from a surface environment similar to that experienced by the Fayetteville meteorite, although their compositions and ages indicate that the gas-rich ordinary chondrites do not come from the moon. There are several differences in detail between chondritic regolith breccias and moon rocks, but these can usually be ascribed to the much greater maturity of lunar regolith breccias.

A major question concerning the regolith breccias is when did the period of regolith activity occur? The meteorites themselves are extremely old—4,600 million years—but were they part of a regolith at that time or did they come from an object in which, like the moon, old materials were being stirred up recently? Several regolith breccias have been taken apart, and it has been found that the formation ages of the individual components cluster around 4,600 million years, even though some of them were shocked or turned partially into glass by regolith activity. This fact suggests that the period of regolith activity was ancient. It may even represent the final stages by which the diffuse dusty material of the early solar system accreted to form chunks of stone and, eventually, the small meteorite parent bodies. On the other hand, a large spectacular black clast of H chondrite was found in the St. Mesmin meteorite, another regolith breccia. The H chondrite clast was heavily shocked 1,400 million years ago but is now embedded in old, unshocked material. At first it seemed that the regolith had been active within the last 1,400 million years, though another possibility is that the H clast was added to an existing consolidated regolith breccia long after the end of regolith activity.

Another piece of evidence regarding the history of the Fayetteville meteorite concerns the interaction of cosmic rays with the meteorite. High energy cosmic rays pass through the meteorite and collide with the atoms it is made of. Pieces break off the atoms producing a whole range of other atoms. Some of these pieces (spallation products) are not normally in the meteorite in very great amounts, and their abundance can be used to estimate the duration of cosmic ray bombardment, provided some means is available for estimating their rate of production. Usually this rate can be calculated from the concentration of radioactive cosmic ray products; their rate of decay at the time of fall is equal to their production rate in space. The cosmic ray exposure ages of some separated clasts from several regolith breccias differ from the majority of the clasts from the same meteorite and appear to have experienced a different exposure to cosmic rays before formation of the breccia. Freshly fallen meteorites are known to have been exposed to cosmic rays immediately prior to their fall to earth because they contain very short-lived radioactive nuclides. Presumably, then, either the regolith is recent, or there have been two episodes of cosmic irradiation: an early one 4,600 years ago during the final stages of accretion (perhaps when the sun passed through an early active [T-Tauri] stage), and one immediately preceding their fall to earth.

Although it seems clear that the regolith breccias are composed of essentially fine-grained dusty dark material into which light clasts of normal material were mixed during a period of regolith activity, it is not clear exactly where the dark material came from. If one assigns a petrologic type on the basis of the parameters described in the preceding chapter, then the light clasts of Fayetteville are type 6 and the matrix is type 4 (Fig. 18). It is normal for regolith breccias to consist of equilibrated clasts in an unequilibrated matrix, implying that the metamorphism responsible for the various petrologic types occurred before or during accretion, if the regolith activity can really be identified with the final stage of accretion. It is contrary to a longstanding view that accretion was an early rapid process which pre-dated metamorphism. An alternative idea is that unequilibrated material was located on the surface of the body, undergoing regolith processes, when an occasional type 6 clast fell on the surface. The type 6 clast may have been excavated from great depths.

Another possible complication is that the dark matrix is not type 3 material at all, but crushed-up light clasts which have been made to mimic the type 3 material by the regolith activity. The continual shock and reworking associated with regolith processing could have caused the heterogeneity. The large amount of volatile substances could have been added from outside, in the form of carbonaceous meteorites crashing onto the surface and solar wind pouring continually onto the surface. The solar wind has the same composition as the sun and carbonaceous meteorites.

One of the most fascinating aspects of regolith breccias is that the dark material sometimes contains recognizable fragments of other meteorites. These fragments have been termed "meteorites within meteorites" by one specialist, while another has described them as "poor man's space probes" because they have brought back to earth samples of space rock we might not otherwise obtain. In most instances the foreign meteorite is an example of the rare CM chondrite class, suggesting that regolith breccias must pass through the region where CM chondrites originate on the way to earth. The Fayetteville meteorite has not been studied well enough for specialists to discover such for-

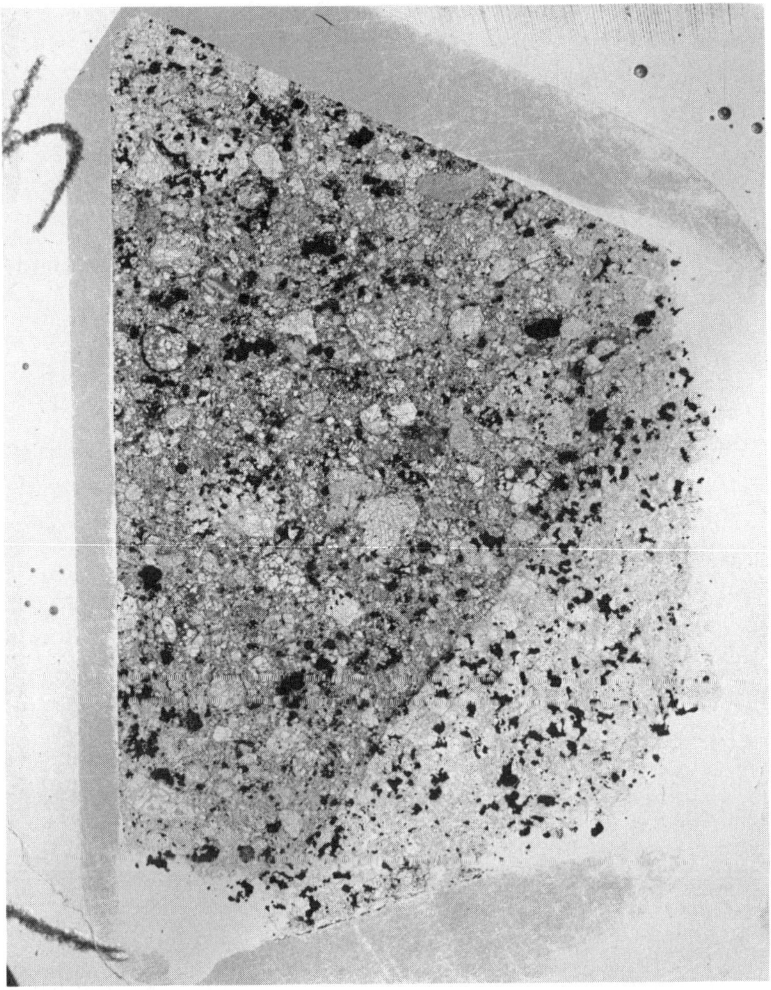

FIGURE 18. Photomicrograph of a thin section of the Fayetteville meteorite showing part of a clast and matrix. Note, in addition to its darker appearance, the sharper and more distinct textures (especially of chondrules) in the dark matrix as compared to the textures of the light clast. The clast is said to be petrologic type 6. (By permission of E. Scott, University of New Mexico)

eigners, but it contains several black inclusions which are possible candidates. The Paragould meteorite is possibly one of the very few to contain foreign fragments which are not CM chondrites; rather, it contains a very small piece of material which some authors have suggested is a melted L chondrite.

The composition of the dark matrix is not identical to that of the light clasts but is enriched in several highly volatile trace elements such as indium, thallium, bismuth, and lead. At first it was thought that this abundance reflected the presence of the CM material because CM chondrites are rich in these volatile elements. If centimeter-sized clasts are present, perhaps millions of sub-millimeter clasts are also present, a condition which would be difficult to detect except by their chemical fingerprint. However, the proportions of these elements are very different from those in the CM chondrites. Either there has been much redistribution of elements destroying the CM proportions, or the matrix material is not a simple mixture of crushed-up equilibrated clasts and CM meteorites.

Clearly, there are many unanswered questions surrounding regolith breccias, and they are currently providing active topics for research—especially the question of the putative relationship between regolith activity and accretion. The Fayetteville meteorite has one of the largest regolith histories known for a meteorite and it seems clear that it will figure prominently in future studies.

VI The Iron Meteorites

The Cabin Creek, Calico Rock, Norfork, Joe Wright Mountain, Sandtown, and Western Arkansas Meteorites

When cut, ground, polished, and etched with a dilute acid (usually 2% nitric acid in alcohol), iron meteorites display a wide variety of internal structures, some of which are most beautiful, all of which originally perplexed their discoverers (Fig. 19). For nearly a century and a half, classification of iron meteorites was based on their internal structure, but in recent years this system of classification has been replaced by one based essentially on bulk composition. Since the structures are unique to meteorites, they are given unique names. Octahedrites comprise the largest group. They have plates of low-nickel alloy, kamacite (symbol α) arranged parallel to the faces of an octahedron with interstices filled with the high-nickel alloy, taenite (symbol γ). The overall texture may be described as fine, medium, or coarse depending on the thickness of the kamacite plates. Hexahedrites are large single crystals of kamacite whose only structural features are thin straight lines arranged along the faces of a cube (hexahedron). The lines are produced by the shock of an impact in space which causes the crystal structure to slip along certain planes. The Calico Rock meteorite is a hexahedrite and owes its brick shape to the meteorite breaking up along these planes. The third major

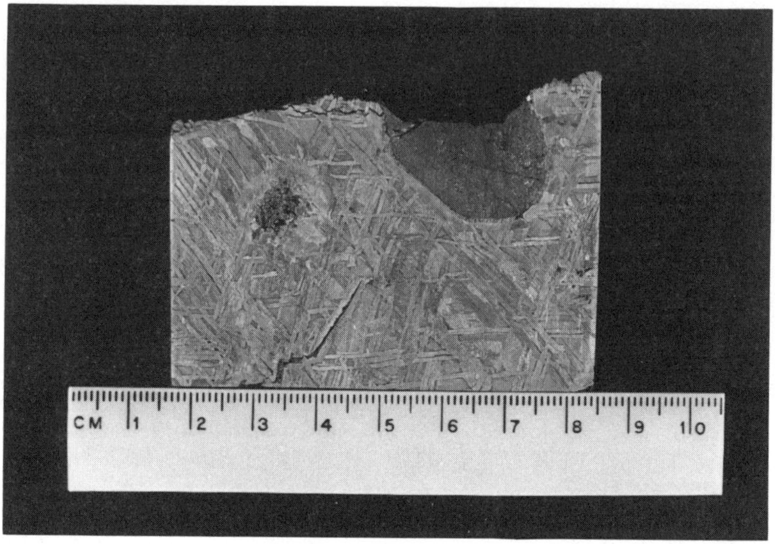

A

B

C

FIGURE 19. Polished sections of three iron meteorites etched with acid to reveal their internal structure: (A) Joe Wright Mountain (*see facing page*) has a medium octahedrite structure with a large troilite nodule (top) and 0.85 mm wide kamacite lathes, (B) Calico Rock (*see facing page*) is a single crystal of kamacite, called a hexahedrite, with no internal structure except fine lines caused by shock impact called Neumann Bands; (C) the Western Arkansas (above) is a fine octahedrite which has been reheated while on earth. It has also been struck intensely, and certain regions seem to have been displaced. (By permission of C. B. Moore, Arizona State University [a], and R. Clarke, Jr., Smithsonian Institution [b and c]).

structural type is the ataxites which appear structureless to the naked eye, but under the microscope are seen to be a fine-grain mixture of kamacite and taenite. In fact, under the microscope the center of the taenite fields in octahedrites is also seen to be a fine-grained mixture of these two phases and is usually called "plessite." Of the seven Arkansas iron meteorites, five are medium octahedrites, one is a hexahedrite, and another is fine octahedrite.

The formation of such structures is well understood in terms of the low-temperature (all-solid) part of the iron-nickel phase diagram. The diagram (Fig. 20) is relatively simple, consisting of α and γ fields separated by a two-phase ($\alpha + \gamma$) field. At high temperatures (>800°C) any alloy of iron and nickel is in the γ phase, but as the temperature falls the alloy crosses into the two-phase region and α precipitates. Until recently it was thought to precipitate as plates, but apparently now it is believed to form rods or needles. The temperature at which the boundary is crossed depends on composition, and there is also some undercooling (perhaps 100°C) because of α nucleation difficulties. If the nickel content is less than about 7%, the alloy falls into the single α phase region at lower temperatures, and a single crystal of α forms. The result is a hexahedrite like Calico Rock. If the nickel is between about 7% and 15%, a three dimensional structure of α plates forms, with residual γ filling the void spaces between the plates. The thickness of these plates, the bandwidth, depends on the nickel content (the more nickel, the more of the high-nickel alloy) and on the rate at which the alloy cooled (faster cooling favors finer structures). The remaining six Arkansas irons are all octahedrites—Western Arkansas with a fine texture, the others with a medium structure. Alloys with >16% nickel pass into the two-phase region at such a low temperature that α formation is severely restricted by diffusion difficulties, and very fine grains or millimeter-sized needles result. There are no examples of such structures in the Arkansas iron meteorites.

The growth of the octahedral structure, named "Widmanstatten structure" after one of its discoverers, has been used to calculate cooling rates. The rate at which the meteorite cooled reflects the size of the object from which it came or burial depth, and a range of cooling rates would indicate a range of burial depths. The results are controversial since they are acutely sen-

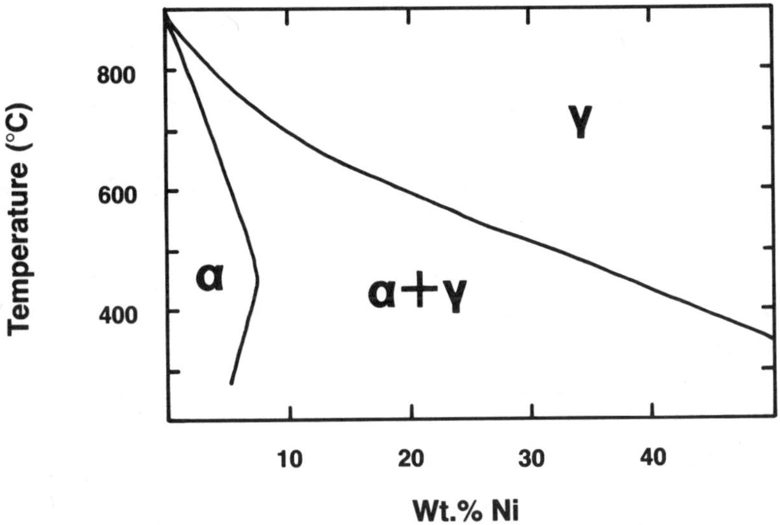

FIGURE 20. The low-temperature half (solid only) of the iron-nickel phase diagram. The structures of most iron meteorites appear to have been fixed by cooling from the γ (taenite, face-centered cubic structure) phase to ~350°C when 5.5% nickel alloys are entirely α (kamacite, body-centered cubic structure); 7–10% nickel alloys are a mixture of lathes of α surrounded by left-over γ; and ~15–20% nickel alloys are a fine-grained mixture of α and γ. (After J. I. Goldstein and R. E. Ogilvie, *Transactions of the Metallurgical Society*, AIME 233, 2083–2087, 1965)

sitive to the assumed diffusion coefficients, and have recently undergone major revision because of new low-temperature diffusion coefficient data and the discovery that kamacite forms as rods rather than plates.

Classification of iron meteorites on the basis of bulk nickel and bandwidth, while in common use for 150 years, is problematic since there are no clear gaps in these properties; there are no clearly separated groups. Although the histogram of nickel contents has three peaks corresponding to hexahedrites, octahedrites, and ataxites, it is not clear that the structural groups reflect really fundamental aspects of the meteorite's history; as we have seen, they reflect to some extent parent body size or burial depth, but not how and where the parent body formed.

Gallium and germanium are two siderophile elements present in irons which do resolve the meteorites into groups, especially when combined with nickel content. Fig. 21 shows the gallium-germanium data with the groups indicated. The available compositional data are presented in Table 7. Calico Rock is a IIA meteorite, Cabin Creek, Joe Wright Mountain, and Norfork are IIIAB, and Western Arkansas is a IVA iron. Sometimes, as in Table 1, the nickel-poor half of groups IIIAB and IIAB are known as IIIA and IIA, respectively, and the nickel-rich halves as IIIB and IIB.

Iron meteorites are not simply iron-nickel alloys, but contain appreciable amounts of cobalt and the minor elements carbon, phosphorus, and sulfur. Cobalt dissolves in the metal phases, usually mainly in the α, but phosphorus and carbon usually also form minor minerals. The carbon forms blebs of cohenite in the center of α plates, or nodules of graphite. Sulfur forms the ubiquitous troilite, iron sulfide. Phosphorus forms a solid solution of Fe_3C and Ni_3C, either as cracked blebs, or occasionally as geometrical shapes called "hieroglyphs." Phosphorus occurs as either schreibersite (solid solution of Fe_3P and Ni_3P) or as small crystals with the same composition but called "rhabdite." Schreibersite occurs on grain boundaries, whereas rhabdites are scattered through the α phase; clearly, they nucleated via different mechanisms.

Buchwald has described the detailed structure of all the Arkansas irons—except Hatfield, of which only a small sample is known, and Cabin Creek, which is apparently too attractive to damage for extensive sampling.[1] The Cabin Creek data in Table 7 are calculated from a section in the center of some small pieces

[1] V. F. Buchwald (1975), *Handbook of Iron Meteorites.*

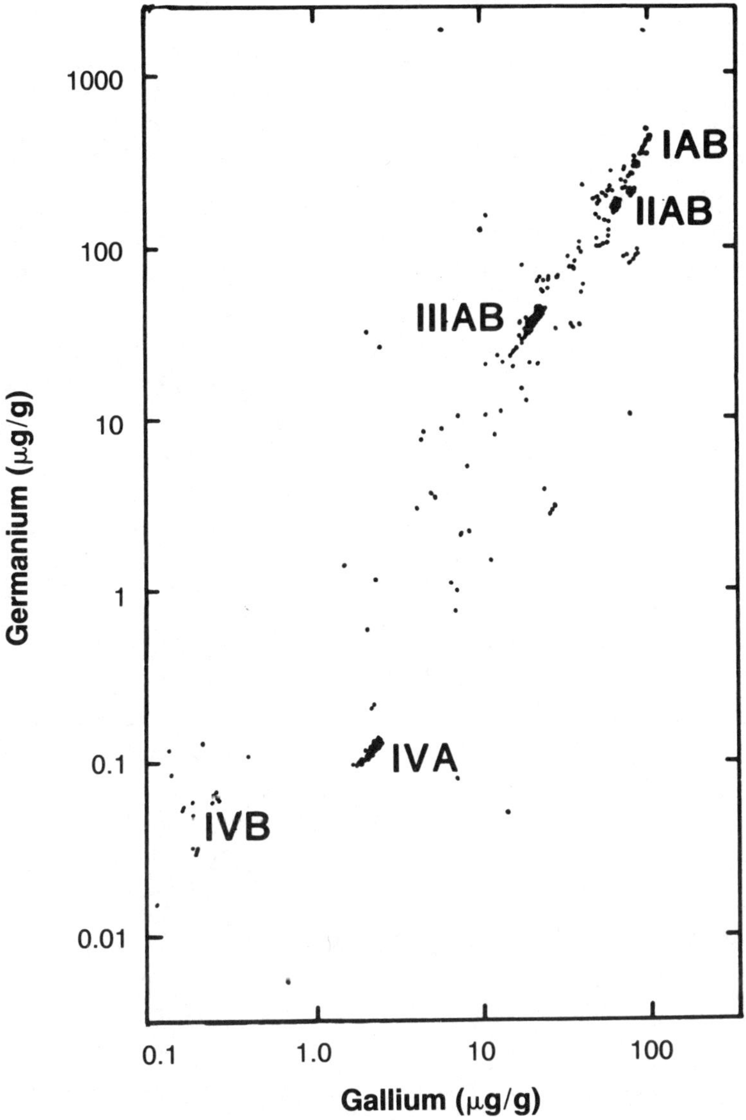

FIGURE 21. Plot of germanium against gallium in iron meteorites (units are parts per million by weight, or micrograms per gram). These two trace elements, together with nickel and sometimes certain mineralogical properties, enable the recognition of at least fifteen separate iron meteorite groups. Several can be resolved on this plot; others require replotting on a larger scale. (After D. W. Sears, *The Nature and Origin of Meteorites*, Oxford University Press, 1979)

taken from the surface, and their original structure has been damaged by the heat of atmospheric passage or by hammering. The Norfork and the Joe Wright Mountain meteorites have normal medium octahedrite structures, although the Neumann Bands are distorted in the latter, presumably because it has experienced an impact in space. An interesting feature of the Joe Wright Mountain meteorite is that superimposed on the normal octahedrite pattern are long thin plates (typically 25 × 5 × 0.05 mm) of schreibersite which are called "Reichenbach lamellae." Apparently they formed from the solid metal before the Widmanstatten plates. There are also sulfide lamellae, of similar dimensions but fewer in number than the phosphide plates, called "Brezina lamellae." They too predate the Widmanstatten structure. Many minor minerals and kamacite tend to be associated with these lamellae.

Trace elements other than gallium and germanium show very broad ranges within all the iron meteorite groups except IAB (and the small IIICD group). Their abundance usually correlates with nickel abundance; sometimes there is a positive correlation (e.g., gold) and sometimes a negative one (e.g., iridium). It appears, then, that the meteorites were produced sequentially during melting-solidification processes, since elements differ in their tendency to distribute themselves between the liquid and solid metal. The simplest process of this sort is called "fraction crystallization" and is well known to geologists.

VII The Violent Cosmos

The Sandtown, Slovak, and Western Arkansas Meteorites

The Sandtown and Western Arkansas, and perhaps Slovak, meteorites bear the marks of another episode in their cosmic history—a violent and catastrophic episode very different from the relatively mild processes of nebula accretion and agglomeration discussed thus far.

Sandtown is a medium octahedrate with the usual Widmanstatten structure of α plates and residual γ fields. The γ has transformed into various forms of plessite containing phosphides and other minor minerals. The meteorite has obviously not been on earth long, for fusion crust is still present in places, and there is a zone 1–4 mm thick around the edges of the meteorite of metal which has been severely heated during atmospheric passage; on most iron meteorites it has been corroded away.

The most interesting feature of this meteorite is that throughout the whole mass of α iron the kamacite has a peculiar texture, as if it had decomposed into numerous smaller grains with a large number of fine γ grains scattered throughout it (Fig. 22). Sometimes the γ grains are lined up. The texture is the result of the meteorite undergoing an intense shock ($>$130 kb) such as would

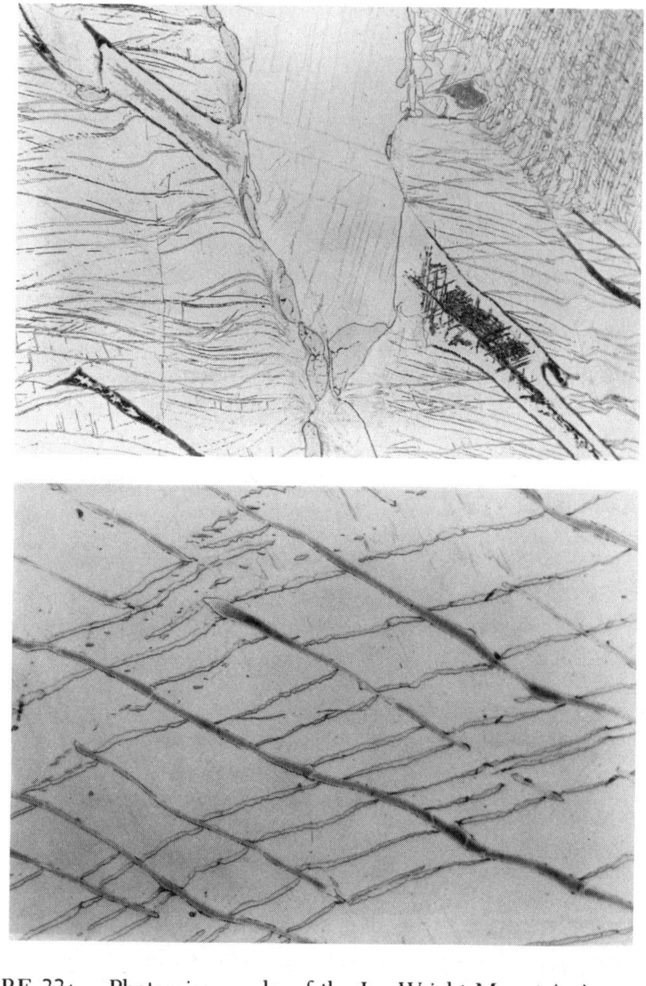

FIGURE 22A. Photomicrographs of the Joe Wright Mountain iron meteorite. In the upper picture are several plates of kamacite (α), each containing Neumann Bands. On either side of the central field of α, running diagonally northwest-southeast, are two fields of taenite (γ) whose centers have decomposed into plessite (fine-grained α, γ mixture). The central α field is surrounded by discontinuous areas of the phosphide, schreibersite. In the lower, higher magnification picture (horizontal field of view 0.28 mm), fine ribbons of γ lie northwest-southeast, while Neumann Bands in the α lie northeast-southwest. (All photomicrographs provided by V. F. Buchwald, Technical University of Denmark, Copenhagen)

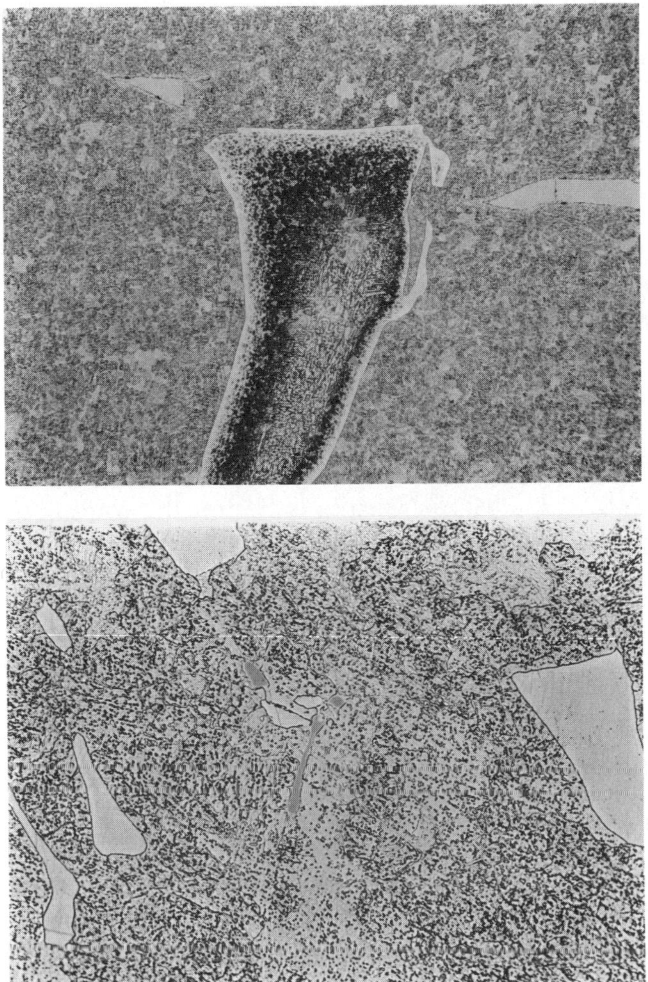

FIGURE 22B. Photomicrographs of the Sandtown iron meteorite. The upper picture (horizontal field of view 1.7 mm) shows a γ field surrounded by α. The structure of both phases was originally much like those in the Joe Wright Mountain, but severe shock, followed by prolonged annealing of ~400°C, caused both phases to take on a different appearance. At higher magnification (lower horizontal field of view 0.22 mm) it can be seen that a great number of fine grains of γ have appeared in the α. (All photomicrographs provided by V. F. Buchwald, Technical University of Denmark, Copenhagen)

be produced by two large objects colliding, and then being held at ~400°C for a long time afterwards.[1]

Western Arkansas is a fine octahedrite whose plates of kamacite have been deformed slightly. It also contains extensive "shear zones," areas where one part of the meteorite has moved with respect to another. These zones can be seen in Figure 19c. These effects are due to some kind of cosmic collision.[2] The kamacite in the meteorite has decomposed to produce a distinctive texture (referred to as α_2) indicating reheating to about 800°C and then fairly rapid cooling; Jain and Lipschutz thought this process was also a result of extraterrestrial events.[3] However, Buchwald found that the weathering products had also been heated to 800°C, indicating that the reheating occurred on earth. He also found chisel and hammer marks, indicating that some of the violence experienced by Western Arkansas was caused by the hand of man. It is not uncommon for iron meteorites to undergo an intense shock event, although only a handful of irons display the effects of subsequent prolonged heating that the Sandtown encountered. Most importantly, nearly all members of the IIIAB class show the shock transformation texture (the so-called ϵ structure), whereas it is absent in classes IAB or the low half of IIAB (sometimes known as IIA). Many of group IVA have been shocked and reheated, whereas many of group IIB have been reheated but not shocked.

The tendency for members of a given iron meteorite class to be either all shocked or all unshocked is strong evidence that the shock process occurred while the members of a class were part of the same parent body. Even stronger evidence for this conclusion—evidence which suggests that it was the shock which actu-

[1] V. F. Buchwald (1975), *Handbook of Iron Meteorites*, 1060.
[2] V. F. Buchwald, 1304.
[3] A. V. Jain and M. E. Lipschutz (1970), *Geochimica Cosmochimica Acta* **31**, 1811.

ally broke up the meteorite parent body—is provided by cosmic ray exposure ages. The cosmic ray exposure ages of iron meteorites range from 90 to 2,275 million years, but most significantly, individual members of group IIIAB tend to have ages around 550–750 million years (Fig. 23). When so many individual meteorites, which have fallen at various times and in various places, have all experienced severe shock and were exposed to cosmic rays at the same time, the conclusion that these meteorites came from a single parent object in space seems inescapable.

Many of the stony meteorites have also undergone shock episodes. The L chondrite class contains many members which are black, instead of the normal pale gray, and contain metal which shows signs of being reheated. They also contain flows of melted sulfide. The shock was violent enough to melt certain silicate minerals, and this can be detected from their physical properties (such as thermoluminescence). It is possible to determine

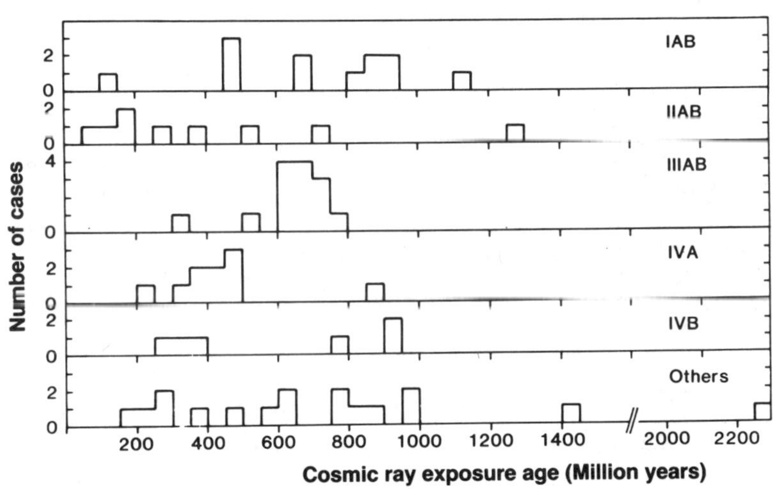

FIGURE 23. Histograms of cosmic ray exposure ages for several iron meteorite classes. (After H. Voshage and H. Feldmann, *Earth and Planetary Science Letters* **45**, 293–308, 1979)

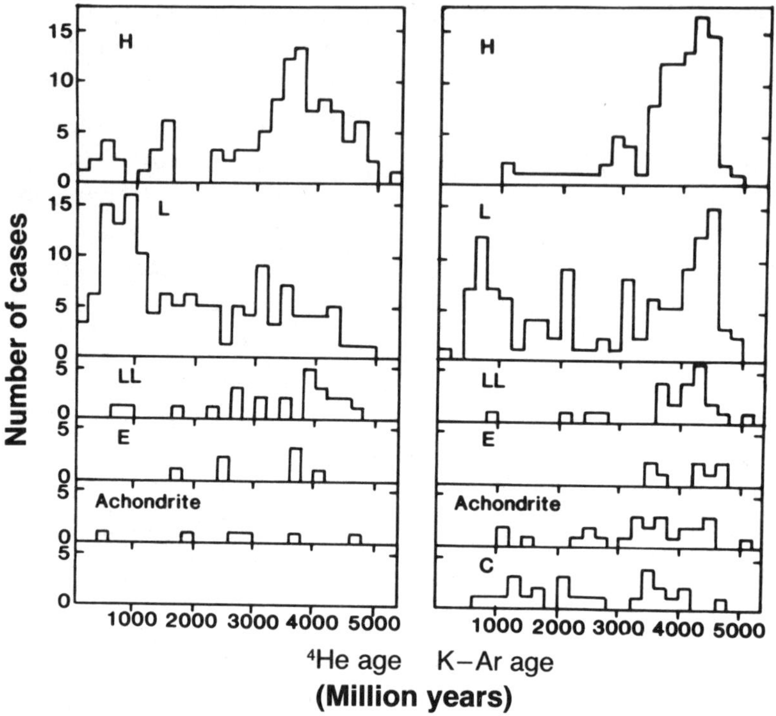

FIGURE 24. Histograms of the ages of stony meteorites calculated from their abundance of argon-40, produced by radioactive decay of potassium-40, and helium, produced by decay of uranium and thorium. The meteorites tend to have ages around 4,600 million years, the age of the solar system, but the L chondrite histogram also has a peak of 500 million years. A major heating event seems to have occurred at that time, resulting in the loss of gases like argon and especially helium. (Data from J. Zahringer, *Geochimica Cosmochimica Acta* **32**, 209–237)

the probable time at which the shock event occurred. Meteorites contain the radioactive species potassium-40 which decays to produce argon-40. Since there are no really important alternative sources of argon-40, it is possible to calculate the time during which the decay has been occurring by measuring the relative amounts of potassium and argon. At the time of shock, the meteorite may have become heated sufficiently for any previously accrued argon to be "boiled-off," so that its potassium-argon "age" is actually the time which has elapsed since the shock event. In recent years the chemical determination of potassium has been replaced with a nuclear method in which potassium is converted to argon-39 by exposure to a neutron flux in the core of a reactor and the argon-39 used to calculate the potassium present. This potassium can be determined simultaneously with the argon-40. Furthermore, since the argon-39 and argon-40 are released from the meteorite by heating, it is sometimes possible to recognize meteorites which were shocked enough to lose easily released gases, but not the most retentive gases. There is a striking tendency for shock-blackened L chondrites to have low potassium-argon ages, generally around 500 million years (Fig. 24). The similarity of this result and the 400–600 million year exposure ages of IIIAB led some authors to speculate that it was a related event—that either both classes came from a single parent body, or that their two parent bodies were involved in a collision.

Cosmic ray exposure ages for stony meteorites are much shorter than for irons; various authors have suggested that this is because they are much more readily broken up, and smaller, more recent, impacts may have been responsible (Fig. 25). In fact, these ages are typically in the range of one to ten million years, comparable with the expected lifetime for objects in Apollo-Amor asteroid orbits. It seems, therefore, that the event which exposed the stony meteorites to cosmic rays may have been the event which sent them onto earth-crossing orbits. However, these ages also point to the conclusion that members of a given class come from a single parent object. The histogram of H chondrite exposure ages has a pronounced peak at four million years. The situation is less clear cut for other classes, although several possibilities exist.

As a rule, the H chondrites are less heavily shocked than the L chondrites, but many of them are black, and the Slovak mete-

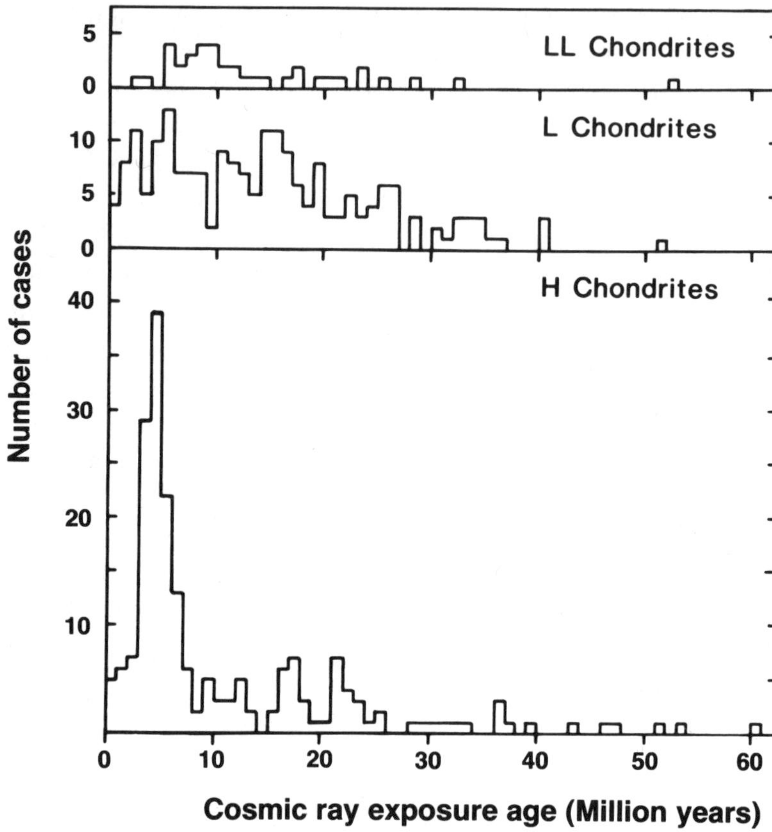

FIGURE 25. Cosmic ray exposure ages for several stony meteorite classes. (After F. Crabb and L. Schultz, *Geochimica Cosmochimica Acta* 45. 2151–2160, 1981)

orite contains large nodules of metal which are commonly regarded as the result of shock. Some authors have suggested that shock should be included in the meteorite's classification; the scheme is presented in Table 8. The mineral olivine shows a number of structural changes as shock pressure increases, while feldspar at lower pressures deforms and at higher pressures turns into a glass called maskelynite. High pressures also produce pock-

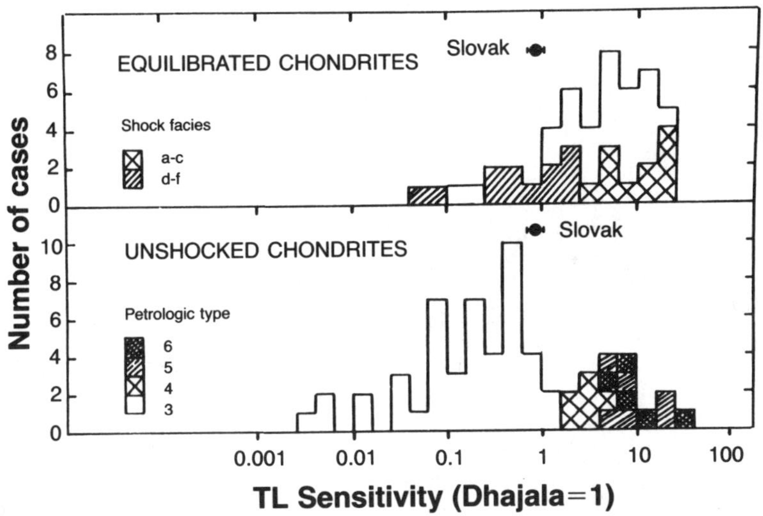

FIGURE 26. Histograms of the TL sensitivity levels in equilibrated (i.e., petrologic type 5 and 6) chondrites of various shock histories and unshocked chondrites of various petrologic types. Type 3 has been subdivided into types 3.0–3.9 on the basis of TL and other data (Table 5). The Slovak meteorite has a TL sensitivity equivalent to type 3.8 ∓ 0.1 or shock facies d–f; petrographic data rule out a petrologic type as low as type 3.8.

ets of melt and the loss of many highly volatile elements, as well as argon-40. The TL sensitivity decreases dramatically at higher shock levels, reflecting the destruction of feldspar, the mineral causing the TL. Thus, a heavily shocked L6 chondrite may sometimes be referred to as an L6f chondrite, while its unshocked equivalent is an L6a. In terms of its textural characteristics, Slovak is shock class b or c,[4] but the thermoluminescence sensitivity is slightly lower than expected for a type 5 chondrite and consistent with shock levels equivalent to shock facies d–f (Fig. 26).

[4]E. R. D. Scott, personal communication, June 1985.

VIII The Pallasite

The Newport Meteorite

The Newport meteorite was found in 1923 but not recognized as a meteorite until it was brought to Harvey Nininger as a result of the publicity he generated over the fall of Paragould. It was originally a 5.6 kg mass, approximately cylindrical in shape. When cut, it was found to have an internal structure very unlike any of the Arkansas meteorites so far described (Fig. 27). It consists of a network of metallic iron-nickel (with 10.7% nickel) which encloses nodules of the silicate olivine (containing about 12 mole % fayalite). The remarkable point is the relative abundance of the two phases; the metal constitutes 56.1 volume percent of the meteorite, while the olivine is 37.0 volume percent. In addition, there is 4.7 volume percent troilite and 2.2 volume percent schreibersite. The metal has undergone conversion into the α and γ phases, and a weakly-formed Widmanstatten pattern exists. The first meteorite discovered with such peculiar properties is known as the Pallas iron, and there are now over forty similar meteorites constituting the pallasite class.

A number of major questions arise concerning the pallasites. It is generally assumed that large amounts of relatively

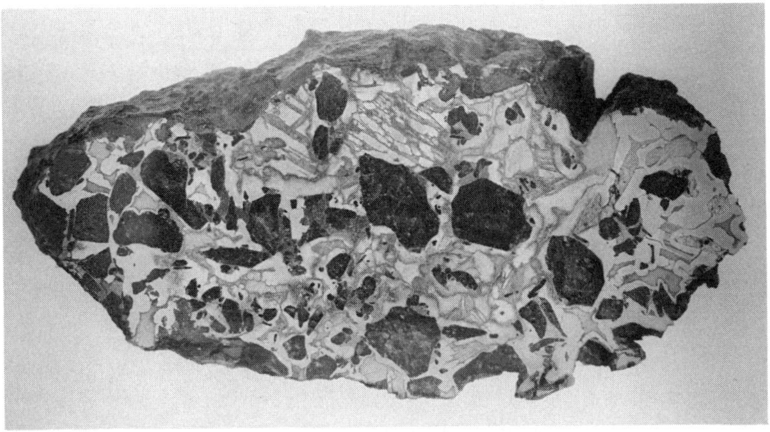

FIGURE 27. Macrosection of the meteorite found at Newport in 1923. It is a member of the rare class of pallasites which consists of a network of metallic iron-nickel alloy (light) enclosing nodules of the silicate olivine (dark gray). The slice is 10.5 cm in length. (By permission of R. Clarke, Smithsonian Institute, and H. H. Nininger)

pure metal can only have been formed by some melting process. Its greater density provides a means of separating metal from the other phases. If the metal which forms the pallasite network entered as a liquid phase, how did the silicates keep from being separated? Furthermore, how did the olivine become separated from the other silicates, and where did the vast quantity of metal come from? Needless to say, the pallasites have been the subject of a great many scientific papers.

With a few exceptions, the metal in the pallasites has a composition very similar to that of the IIIAB iron meteorites, and it is commonly assumed that these two classes are related. An abundance of nickel, gallium, and germanium—elements of particular significance for iron meteorite origins—is present in both classes. But other elements such as gold, arsenic, and chromium, which show enormous spread in the IIIAB irons, put the pallasites at one end of the IIIAB range, indicating that their origin is related to, but not identical to, that of the IIIAB class.

The composition of the olivine also provides some clue as to the origin and history of the meteorites. The amount of iron in the olivine is intermediate between that found in chondrite meteorites and that found in the few silicate inclusions in IAB iron meteorites. This evidence suggests a separate formation environment from that of the other classes.

Pallasite olivine is also noted for its high nickel content, implying that the exchange of nickel between metal and silicates ceased at high temperatures; at lower temperatures, the low nickel content observed in other chondrite clases is predicted theoretically. Olivine is the first of the major silicates to crystallize out of a melt with chondritic composition, so that if a chrondritic mass were melted it would form and fall to the bottom of the magma chamber to produce what is known as an "olivine cumulate," resulting in relatively pure olivine. Various authors have suggested that the magma chamber was located on a molten metal core, and that the pallasites are samples of the core-mantle boundary. The influx of metal into the olivine cumulate must have been rapid and it must have cooled quickly in order not to separate again. After being mixed, some of the pallasites underwent a long annealing so that their olivine grains became rounded—a surface tension effect. The Newport meteorite is intermediate between those meteorites with highly angular grains and those with highly rounded grains.

IX Origins

Strange as it may seem, meteorite orbits tell us relatively little about the formation location of meteorites because the orbits by which they came to earth are highly unstable and could not have existed very long (see Chapter II). The meteorites were recently put into these orbits, either by impact or by the gentle pushing and pulling of the gravitational fields of other objects, probably planets.

To a first approximation, the age, composition, and texture of the chondritic meteorites seem to indicate a fairly simple origin at the earliest epochs in the history of the solar system. Chondrites do not resemble rocks which would be found on the earth, moon, or any other large planetary object; photographs from satellites indicate that all the planets have undergone planet-wide melting, in some cases to produce a metallic core and silicate mantle and in others to produce major lava flows. If we look at objects in the present solar system for possible meteorite parent bodies, we are left only with asteroids (chunks of rock a few meters to a few hundred kilometers in size which orbit between Mars and Jupiter) and comets, which are dusty "snow-balls," the "snow" being a mixture of frozen water, methane, ammonia, and carbon dioxide. The solid stony residues of comets may become "asteroids" on orbits which cross the earth's orbit.

It has been argued that the abundance of solar wind gases in meteorites is too high for a cometary origin; comets spend most of the time in deep space, too far removed from the sun to trap gases. Some evidence also exists linking certain of the meteorites to the asteroids. If the reflected sunlight from an asteroid is passed through a spectrometer a spectrum is produced which reflects the surface composition of the asteroid and which can be compared with similar spectra for meteorites (Fig. 28). There are some remarkable matches: the achondrites and carbonaceous chondrites, and probably irons and stony irons, may well have been associated with parent objects in the asteroid belt. The most perplexing aspect of such studies, however, is that most of the asteroids produce a spectrum matching that of rather rare meteorites in the world's collections. At the same time, the most abundant meteorite class, the ordinary chondrites, does not have a match among the main-belt asteroids. Since most authors believe that ordinary chondrites come from asteroids, it has been suggested that some process, perhaps space erosion, has caused changes in the surface characteristics which are masking the true asteroid spectra.

For many years, though, it seemed that the major difficulty in the hypothesis that meteorites came from the asteroid belt was not based on reflectivity spectra, but rather on orbital mechanics; it seemed that there was no mechanism for providing the change in orbit required to send a main belt asteroid to earth. In recent years, however, means have been discovered involving the influence of Jupiter's gravitational field, though they do not appear efficient enough to provide the observed rate of meteorite fall on earth. Nevertheless, despite a few exceptions mentioned below, there seems to be a consensus that meteorites come from the asteroid belt.

A third, little-discussed possibility is that instead of coming entirely from either comets or the asteroid belt, the meteorites

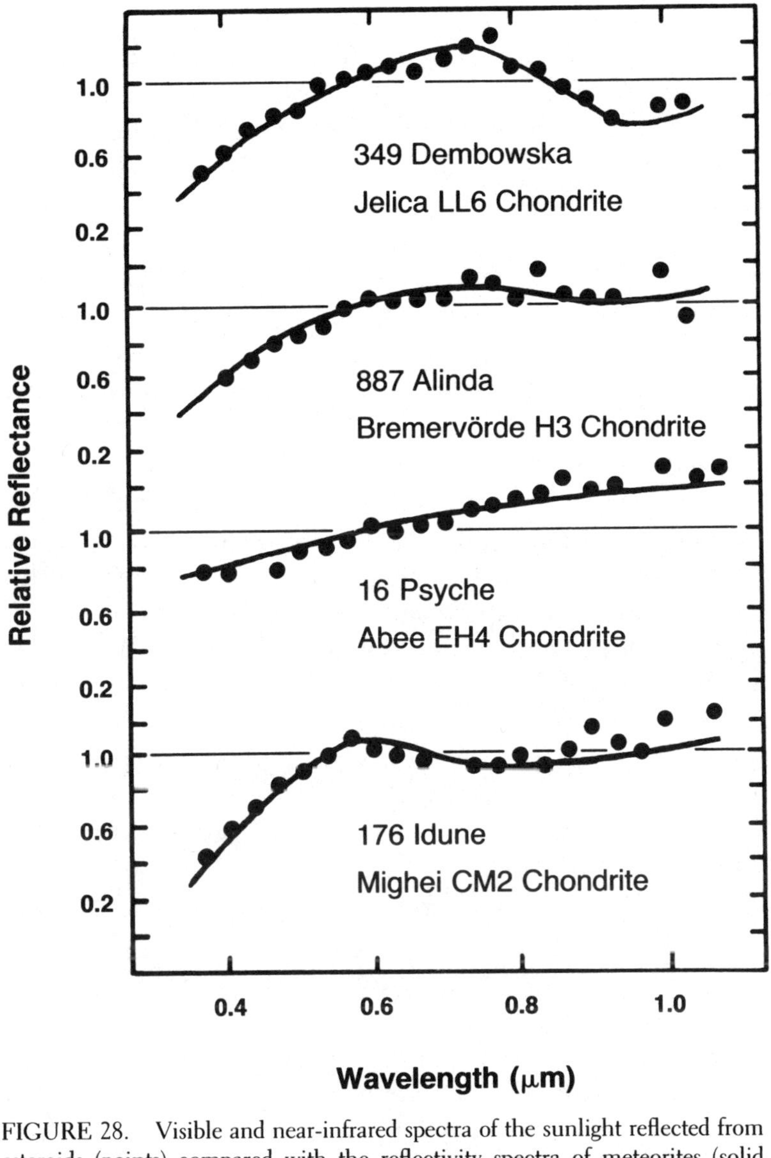

FIGURE 28. Visible and near-infrared spectra of the sunlight reflected from asteroids (points) compared with the reflectivity spectra of meteorites (solid curves). (After C. R. Chapman, *Geochimica Cosmochimica Acta* **40**. 701–719, 1976)

originate from a wide variety of additional sources. They may well also contain some material from outside the solar system. Evidence for this conclusion is provided by the meteorites themselves. Most authors believe that they formed by a process related to the formation of the sun and planets. These meteorites may almost be considered by-products of planet formation, for they are the material which gathered together the dust remaining after the formation of the solar system. Within the last few years two mysteries have been investigated: where did the dust come from, and where did the components of the meteorite come from? It has become evident that the components did not have a single origin.

It is quite likely that before the formation of the sun, planets, their satellites, and the meteorites, there existed a large swirling mass of gas and dust similar to, but very much smaller than, the nebulae observed by astronomers. This postulated mass of gas and dust has become known as the primordial solar nebula. The hypothesis is that as the sun began to form, the nebula collapsed and became heated so that in the immediate vicinity of the sun everything became gaseous. Through laboratory data on chemical reactions, it is possible to calculate what happened when the gas cooled down to current space temperatures (Fig. 29). A pressure for the nebula has to be assumed; in Figure 29, 10^{-6} atmosphere was taken, but the results are not too sensitive to the value chosen. The only major effect of pressure is that at pressures above 10^{-3} atmosphere, metal condenses ahead of olivine and pyroxene, whereas at lower pressures silicates condense at higher temperatures than the metal. At about 1230 K, and at 10^{-6} atmosphere, the major minerals, olivine, pyroxene, and feldspar form; and at about 1200 K, metal forms. Between 600 and 700 K hydrogen sulfide in the gas reacts with some of the metal to form iron sulfide, and between 500 and 600 K, more of the metal reacts with H_2O in the gas to oxidize the iron and form the iron olivine and pyroxene. At lower temperatures still, magnetite

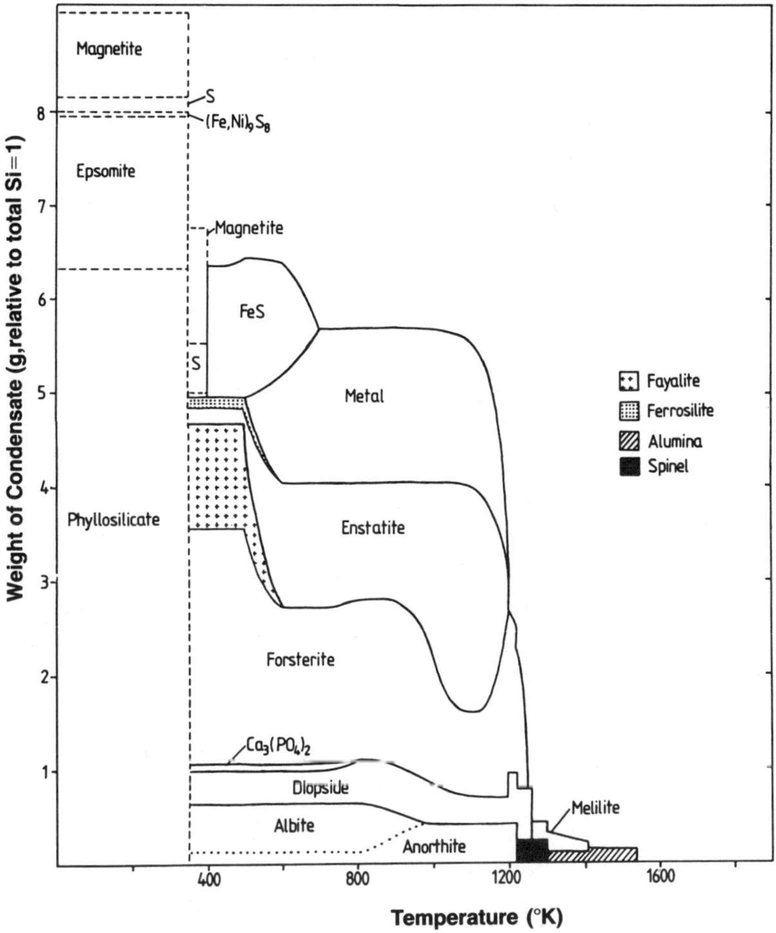

FIGURE 29. The calculated solid components to form in a gas with solar composition at various temperatures. (The results below 400°K are guesses based partly on calculations and partly on the observed mineralogy of CI chondrites.) The weight of each mineral is expressed in terms of 1 gram of silicon in the total system. Formulae for the minerals are: alumina (Al_2O_3), melilite ($Ca_2MgSiO_2O_7$), spinel ($MgAl_2O_4$), anorthite ($CaAl_2Si_2O_8$), albite ($NaAlSi_3O_8$), diapside ($CaMySI_2O_6$), forsterite (Mg_2SiO_4), enstatite ($MgSiO_3$), metal (Fe,Ni), fayalite (Fe_2SiO_4), ferrosilite ($FeSiO_3$), magnetite (Fe_3O_4), phyllosilicate (poorly known, probably $Mg_3Si_2O_5[OH]_4$), epsomite ($CaSO_4$). (After D. W. Sears, *The Nature and Origin of Meteorites*, Oxford University Press, New York, 1979)

(Fe_3O_4) and carbonaceous compounds form. A range of oxidation states similar to that observed in chondrites (Fig. 11) could result from the agglomerating of material over a range of temperatures and pressures. Furthermore, the variations in the magnesium to silicon and calcium to silicon ratios, which distinguish the classes (Fig. 12), could result from the removal of calcium- and magnesium-rich phases which form at high temperatures. Thus, it is thought that the enstatite chondrites, which are highly reduced and have low calcium-to-silicon ratios, formed nearest to the sun, and that the highly oxidized high meteorites with calcium-to-silicon ratios (i.e., the carbonaceous chondrites) formed farthest from the sun. Intermediates in these properties, and presumably formation locations, are the ordinary chondrites. Very similar calculations have been used to explain why the oxidation state and calcium-to-silicate ratio of the planets and satellites vary systematically with distance from the sun.

Any attempt to be more precise about formation distances is a somewhat speculative business. Some authors place the enstatite chondrites on the inner fringes of the asteroid belt and the carbonaceous chondrites on the outer fringes, while others, the present writer included, place the enstatite chondrites much closer to the sun than the earth, perhaps near the orbit of Mercury, with the carbonaceous chondrites in the asteroid belt and the ordinary chondrites at about earth distance.

The iron meteorite compositions—in particular their gallium, germanium, and nickel abundance patterns—can also be explained in terms of such chemical calculations, but again it must be assumed that a wide range of distance from the sun is involved. The theory described in Figure 30 places the formation location of the IAB meteorite body at earth distance from the sun, the IIIAB and IVB bodies in the asteroid belt, and IIA near the orbit of Saturn or Jupiter.

The main difficulty with the hypothesis that meteorites originated all over the solar system before being stored in the as-

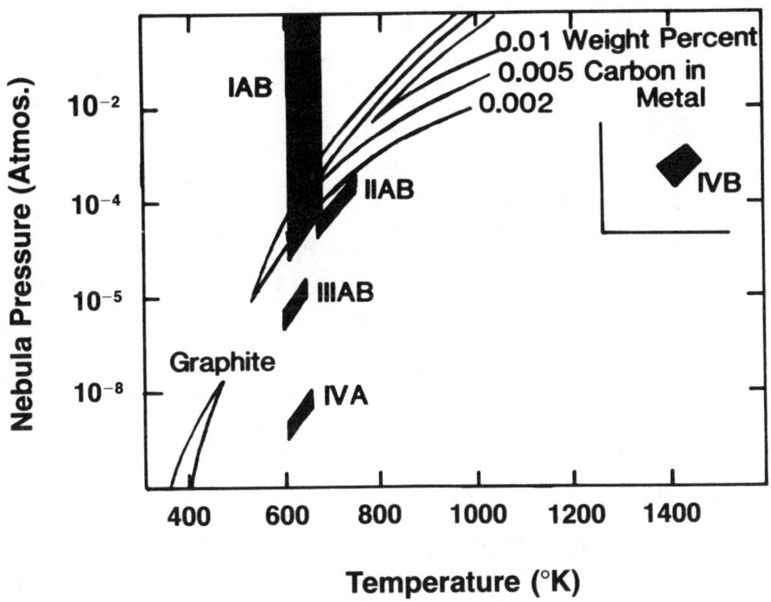

FIGURE 30. Graphical presentation of an idea for the formation of the major iron meteorite classes. The boxes locate pressures and temperatures at which appropriate gallium, germanium, and nickel values are found in the metal; these elements define the classes. The curved V-shaped contours map out the pressure-temperature conditions in which are produced the indicated carbon concentrations in the metal, explaining the carbon variations between the groups. Below 700K, sulfide should occur in the solid phase, explaining its presence in all classes but IVB. These calculations are consistent with IAB forming slightly closer to the sun than IIAB, and much closer than IIIAB and IVB (but these two at very different temperatures) and very much closer than IVA. (After D. W. Sears, *Earth and Planetary Science Letters* **41**, 128–138, 1978)

teroid belt involves the orbital mechanics. The laws governing planetary and other orbits are well known, but the mechanics of orbit change have proved remarkably difficult to explain. For some time it was thought that there was no way to remove material from the asteroid belt, but methods, albeit of low efficiency, have been found. It was also thought that meteorites could not come from planetary-sized objects, but a very small achondrite group known as the "SNC meteorites" after Shergottites, Nahklites, and Chassigny, may well be from Mars. They are all igneous rocks which melted 1,500 million years ago and which contain trapped gases resembling those in the Martian atmosphere. Other achondritic meteorites from Antarctica—Allan Hills A81005, Yamato 791197, and Yamato 82192—have the texture, bulk composition, and formation age corresponding to lunar regolith breccias; there is little doubt they came from the moon. While it is clear on chemical and structural grounds that chondrites do not come from planetary-sized objects, the SNC and lunar meteorites demonstrate that a method exists, probably impact, for producing major orbit changes which do not destroy the objects.

Whatever the formation location of the various meteorite classes, it is impossible to explain all of the properties of primitive meteorites in terms of the gathering together of dust at a certain pressure and temperature. Several properties of several meteorite classes seem to require that grains or small clasts of material which formed under very different circumstances have been mixed together to produce the meteorites. Both the enstatite chondrites and the carbonaceous chondrites appear to be mixtures of high-temperature agglomerates and low-temperature material. The ordinary chondrites, too, are mixtures. The most primitive members of the class contain metal grains with a small quantity of metallic silicon. Silicon-bearing metal can form only at high temperatures before the major silicates form to compete for the silicon. The same meteorites contain mixtures of graphite and magnetite that require very low temperatures.

The most exotic possibility of all is that certain meteorites may contain material from outside the solar system. It has become an axiom of modern science that the ratios of the different isotopes of a given element are constant unless altered by nuclear processes (such as radioactive decay and interaction with cosmic rays) or so-called mass-dependent processes. Both nuclear processes and mass-dependent processes are well understood; nuclear processes can be studied in the laboratory and can be tabulated, and mass-dependent processes can be recognized for the way different masses of a given element have been affected. In recent years a great many instances of unusual isotope ratios, called "isotope anomalies," have been found in these meteorites which are not due to these mechanisms. Probably the best known of such cases is that of oxygen (see Fig. 14). Certain high-temperature aggregates in carbonaceous meteorites contain excess amounts of the light isotope 0–16. Some have suggested that high-energy nuclear reactions in the early solar system were responsible, and recently 0–16 enrichments have been made in the laboratory using electric discharges in oxygen-bearing gases. However, much thought has been given to the possibility that the 0–16 exists in grains which came from supernova ejecta; there is a certain layer in supernova atmospheres enriched in 0–16. Enrichments in other isotopes have been interpreted similarly. Instead of being totally gaseous, the primordial nebula from which the sun and planets formed must have contained grains of solid matter. Some scientists have even argued that most of the solid material in the early solar system was never heated very much, and that a nebula heated entirely above 2000°K is astrophysically unlikely.

It has already been noted that formation ages for all the chondritic meteorites are 4,600 million years, the age of the solar system. In addition to the discovery of numerous isotopic anomalies in meteorites, the recent flurry of isotope measurements has also resulted in the discovery of many isotope enrichments which are the result of the decay of elements which ceased to exist soon after the formation of the solar system. Aluminum-26, for example, will decay away in several million years, but the product of its decay, magnesium-26, is stable and will remain in the meteorite indefinitely as a "ghost" of the aluminum-26. Evidence for other extinct nuclides has been found, notably for plutonium-244 and iodine-129. Measurements of extinct isotopes have led to two conclusions. First, the solar system formed soon after the elements formed, since they would have decayed away in a very short time. The elements formed inside stars and were ejected into space through a supernova explosion. It has been speculated that since the solar system formed so soon after the end of element formation the two events were related and that the supernova explosion may have triggered the formation of the solar system. Second, since the abundance of the decay products, when it can be determined, is fairly similar in all the meteorites examined, the meteorites must have formed over a narrow time interval of within twenty million years. The process for converting the primordial solar nebula, whether it was hot or cold, was apparently a very rapid one.

The study of meteorites is an exciting, complex (because of its multidisciplinary character), and at times frustratingly speculative business, and it commands a vast literature. In the present small work I have tried to provide a view of the subject through the window of Arkansas meteorites. As a result, some topics may have been overemphasized, while others may have been treated too briefly. For deeper and more technical coverages many books and articles are available, some of which are listed in Appendix 1.

Appendices

TABLE 1

The Thirteen Arkansas Meteorites: Summary Data

Meteorite and County	Date of Fall or Find	Coordinates*	Mass (kg)	Class† and Structure	References‡
Cabin Creek, Johnson Co.	Fell 3:17 P.M., Mar. 27, 1886	35°27′N, 93°19′W	48.2	IIA iron, structurally O_m	1, 2, 39
Calico Rock, Izard Co.	Found 1938	35°6′N, 92°9′W	7.28	IIA iron, structurally H	2, 3, 4, 5, 38
Fayetteville, Washington Co.	Fell 11:58 A.M., Dec. 26, 1934	36°3′N, 94°10′W	2.28	H chondrite, regolith breccia	6, 7, 8, 36, 37
Hatfield, Polk Co.	Found 1941	34°29′N, 94°27′W	0.021	class unknown, structurally O_m	2, 9
Joe Wright Mtn., Independence Co.	Found 1884	35°49′N, 91°32′W	42.7	IIIB iron, structurally O_m	2, 5, 10, 11
Miller, Cleburn Co.	Fell 9:00 A.M., July 13, 1930	35°24′N, 91°16′W	16.7	H5 chondrite	12, 13
Newport, Jackson Co.	Found 1923	35°36′N, 95°16′W	5.6	P	14, 35

Norfork, Baxter Co.	Fell Oct. 1918	36°13′N, 92°17′W	1.05	IIIA iron, structurally O_m	2, 5, 9, 15, 16, 17, 18, 19, 20, 21, 22, 23, 38
Paragould, Green Co.	Fell 4:08 A.M., Feb. 17, 1930	36°4′N, 90°30′W	408	LL5 chondrite	24, 25, 26, 27, 28, 39, 40
Sandtown, Independence Co.	Found 1938	35°56′N, 91°38′W	9.4	IIIA iron, structurally O_m	2
Slovak, Prairie Co.	Found early 1960s	34°38′N, 91°32′W	8.22	H5	this work
Success, Clay Co.	Fell Apr. 18, 1924	36°29′N, 90°40′W	3.5	L6 chondrite	30, 31
Western Arkansas, Montgomery Co.	Found before 1890	34°30′N, 93°30′W	1.75	IVA iron, structurally O_f	2, 32, 33, 34

TABLE 1 *(cont.)*

*The most recent values given by M. Hey (1966), *Catalog of Meteorites*, British Museum (Natural History) London; B. Mason (1962), *Meteorites*; J. Wiley, New York or V. F. Buchwald (1975), *Handbook of Iron Meteorites*.

† IIIA, IIIB, and IVA irons and H and L chondrites are defined in the text. O_m, O_f, and P refer to medium octahedrites, fine octahedrites, and pallasites, respectively.

‡ References:

1. G. F. Kunz (1887), *American Journal of Science* **33**, 494–501.

2. V. F. Buchwald (1975), *Handbook of Iron Meteorites*. Berkeley: University of California Press.

3. *Meteoritical Bulletin* (1965) No. 33.

4. J. T. Wasson (1969), *Geochimica Cosmochimica Acta* **33**, 859–876.

5. C. B. Moore, C. F. Lewis, and D. Nava (1969) in *Meteorite Research* (P. M. Millman, ed.) 738–748.

6. D. P. Richardson (1935), *Popular Astronomy* **43**, 384–385.

7. E. R. D. Scott and R. S. Rajan (1981), *Geochimica Cosmochimica Acta* **45**, 53–67.

8. L. Schultz and H. Ruse (1983) *Helium, neon and argon*. A data compilation, Max-Planck-Institut für Chemie, Mainz. (Compilation of the numerous light inert gas measurements made on the Fayetteville.)

9. H. Nininger and A. D. Nininger (1950), The Nininger Collection of Meteorites, American Meteorite Laboratory, Winslow, Ariz.

10. W. E. Hidden (1886), *American Journal of Science* **31**, 460–465.

11. E. R. D. Scott, J. T. Wasson, and V. F. Buchwald (1973), *Geochimica Cosmochimica Acta* **37**, 1957–1983.

12. C. A. Reed (1931), *Natural History* **31**, 109.

13. B. Mason and H. B. Wiik (1961), *Geochimica Cosmochimica Acta* **21**, 266–271.

14. H. H. Nininger (1932), *American Journal of Science* **23**, 78–80.

15. J. T. Wasson and J. Kimberlin (1967), *Geochimica Cosmochimica Acta* **31**, 2065–2098.

16. H. H. Nininger (1937), *Popular Astronomy* **45**, 449–454.

17. L. LaPaz (1938), *Popular Astronomy* **46**, 523–524.

18. R. R. Jaeger and M. E. Lipschutz (1967), *Geochimica Cosmochimica Acta* **31**, 1811–1832.

19. E. L. Fireman and D. Schwarzer (1957), *Geochimica Cosmochimica Acta* **11**, 252–262.

20. D. Tilles and M. S. Tamers (1963), *Nature* **200**, 563–565; **201**, 808–809.

21. O. A. Schaeffer and D. Heymann (1965), *Journal of Geophysical Research* **70,** 215–224.

22. H. Voshage (1967), *Zeitshchrift Naturforschung* 17a, 422–432.

23. H. Hintenberger, L. Schultz, H. Waenke, and H. Weber (1967), *Zeitschrift Naturforsch* 22A, 780.

24. C. C. Wylie (1931), *Scientific American* **144,** 180–181.

25. C. C. Wylie (1930), *Popular Astronomy* **38,** 246–247, 308–309, 387–392.

26. C. C. Wylie (1930) *Science* **72,** 66–68.

27. H. E. Nelson and W. J. Thomsen (1947), *Popular Astronomy* **55,** 448–452.

28. H. H. Nininger (1950), *Out of the Sky,* Dover Publications.

29. W. R. Van Schmus and J. A. Wood (1967), *Geochimica Cosmochimica Acta* **31,** 747–766.

30. B. Mason (1965), *Geochimica Cosmochimica Acta* **27,** 1011–1023.

31. *Meteoritical Bulletin* (1957) No. 5.

32. G. P. Merrill (1927), *Proceedings of U.S. National Museum* 72, art 4, 1–4.

33. R. Schaudy, J. T. Wasson, and V. F. Buchwald (1972), *Icarus* **17,** 174–192.

34. A. V. Jain and M. E. Lipschutz (1970), *Geochimica Cosmochimica Acta* **34,** 883–892.

35. E. R. D. Scott (1977), *Geochimica Cosmochimica Acta* **41,** 349–360.

36. R. Rider and H. Waenke (1969) in *Meteorite Research* (P. M. Millman, ed.) 75–85.

37. W. Nichiporuk, A. Chodos, E. Helin, and H. Brown (1967), *Geochimica Cosmochimica Acta* **31,** 1911–1930.

38. H. Voshage and H. Feldman (1979), *Earth Planetary Science Letters* **45,** 293–308.

39. A. Kracher, J. Willis, and J. T. Wasson (1980), *Geochimica Cosmochimica Acta* **44,** 773–787.

40. K. Fredriksson, J. Nelen, and B. J. Fredriksson (1968) in *Origin and Distribution of the Elements* (L. H. Ahrens, ed.), 457–470.

TABLE 2

Major Meteorite Classes

Chondrites		Number Known	Fall Frequency (%)
Enstatite chondrites	EH	14	0.60
	EL	10	0.72
Ordinary chondrites*	H	623	33.3
	L	605	38.4
	LL	129	8.0
Carbonaceous chondrites	CI	5	0.60
	CM	23	2.2
	CO	13	0.60
	CV	15	0.72
Ca-rich achondrites			
Howardites		25	2.2
Eucrites		46	3.0
SNC		6	0.60
Ca-poor achondrites			
Aubrites		27	1.1
Diogenites		18	1.1
Ureilites		19	0.48
Stony-irons			
Mesosiderites		30	0.72
Pallasites		38	0.36
Irons	IAB	105	0.75
	IIAB	71	0.45
	IIIAB	197	1.42
	IVA	55	0.39
	IVB	12	0.09

Sources: A. L. Graham, A. W. R. Beran, and R. Hutchison (1985), *Catalog of Meteorites*, 4th ed., Univ. of Arizona Press, 460, with modification.

*Excluding Antarctic meteorites, since pairing may be a significant problem.

TABLE 3

Bulk Compositional Data of Arkansas Stone Meteorites*

	Fayetteville L	Fayetteville D	Miller	Paragould	Slovak
Na mg/g	---	---	6.0	4.7	4.6
Mg mg/g	---	---	142.9	141	140
Al mg/g	---	---	10.8	12.2	13.1
Si mg/g	---	---	178.1	206.2	---
P mg/g	---	---	2.2	---	---
S mg/g	---	---	19.7	---	---
K mg/g	---	---	0.8	3.1	7.1
Ca mg/g	---	---	11.5	15.9	---
Ti mg/g	---	---	0.85	---	---
V μg/g	---	---	---	---	83.9
Cr mg/g	---	3.7	2.2	---	3.5
Mn mg/g	2.3	2.4	2.4	---	2.3
Fe mg/g	---	---	270.1	198	231
Co mg/g	---	0.993	0.610	---	0.694
Ni mg/g	---	---	16.0	14.3	18.6
Zn μg/g	---	---	---	---	65.3
Ga μg/g	5.54	5.30	---	---	9.6
Ge μg/g	10	13	---	---	---
As μg/g	---	---	---	---	2.3
Se μg/g	---	---	---	---	8.81
Pd μg/g	---	1.15	---	---	---
Im μg/g	1.1	9.1	---	---	---
Sm μg/g	---	---	---	---	318
Eu μg/g	---	79	---	---	146
W μg/g	---	220	---	---	---
Re μg/g	---	210	---	---	---
Ir μg/g		0.95	---	---	0.78
Ar μg/g	210	220	---	---	205

Source: Data taken from Refs. 13, 25, 36, 37 in Table 1.

*No data available for the Success meteorite. Fayetteville L and D refer to the light clasts and dark matrix respectively (see Fig. 16). The concentration units are milligrams per gram (mg/g, or parts per thousand) and micrograms per gram (μg/g, or parts per million).

TABLE 4

Mineral Abundances in the Miller Meteorite Calculated from Its Bulk Composition

Mineral	Formula	Percentage
Metal	Fe,Ni	21.6
Troilite	FeS	5.4
Olivine	$(Mg,Fe)_2SiO_4$*	18.9
Pyroxene	$(Mg,Fe)SiO_3$†	39.8
Diopside	$CaMgSi_2O_6$	2.4
Albite‡	$NaAlSi_3O_8$	6.9
Anorthite‡	$CaAl_2Si_2O_8$	1.6
Orthoclase‡	$KAlSi_3O_8$	0.6
Apatite	$Ca_5(PO_4)_3(OH,F,Cl)$	1.2
Chromite	$FeCr_2O_4$	0.5
Ilmenite	$FeTiO_3$	0.3

Source: Data taken from Ref. 13 in Table 1.

*Indicates a solid solution of Mg_2SiO_4 (forsterite) and Fe_2SiO_4 (fayalite).

†Indicates a solid solution of $MgSiO_3$ (enstatite) and $FeSiO_3$ (ferrosilite).

‡Albite, anorthite, and orthoclase actually occur as a single solid solution, called feldspar.

TABLE 5

Petrologic Types of Chondritic Meteorites*

Property	Petrologic Type									
	1	2	3.0–3.1	3.2–3.3	3.4–3.5	3.6–3.7	3.8–3.9	4	5	6
Homogeneity of Olivine†	---	——— ≥ 50 ———→			40–50	20–40	5–20	<5	Uniform	
Structural state of low Ca pyroxene	---	Predominantly monoclinic						>20% mono-clinic	<20% mono-clinic	Ortho-rhombic
Feldspar	---	Absent ———→ Absent or rare						<2 μm grains	<50 μm grains	>50 μm grains
Primary glass	---	Clear isotropic ———→ Turbid						Turbid if present	Absent	
Thermoluminescence sensitivity (Dhajala = 1)	---	---	<0.01	.010–.046	.046–.22	.22–1.0	1.0–4.6	1–10	>5	

* There are a few type 4–6 chondrites whose petrologic type definitions differ slightly from these. The definitions above apply especially to the ordinary chondrites.

† Standard deviation of the mole % fayalite in the olivine divided by the mean fayalite expressed as a percentage.

Table 5 (cont.)

TABLE 5 *(cont.)*

Property	Petrologic Type									
	1	2	3.0–3.1	3.2–3.3	3.4–3.5	3.6–3.7	3.8–3.9	4	5	6
Heterogeneity of metal‡	---	---	>17	10–17	6.0–10	2.5–6.0	<2.5		Uniform	
Average Ni content of sulfide minerals	---	>0.5%	<0.5%							
Chondrule delineation	no chondrules		very sharply defined					well defined	readily defined	poorly defined
Matrix texture	all fine-grained opaque	much opaque	% Transparent microcrystalline <20	10–20	~20–~50	>60		100	Recrystallized matrix	
Matrix $\left(\frac{FeO}{FeO+MgO}\right)^{\S}$	---	---	>1.7	1.5–1.7	1.3–1.5	1.3–1.3	<1.1	---	---	

Sources: W. R. Van Schmus and J. A. Wood (1967), *Geochimica Cosmochimica Acta* **31**, 747–765. D. W. Sears, J. N. Grossman, C. L. Melcher, L. M. Ross, and A. A. Mills (1980), *Nature* **287**, 791–795; G. R. Huss, K. Keil, and G. J. Taylor (1981), *Geochimica Cosmochimica Acta* **45**, 33–51.

‡ Standard deviation of the nickel content in the kamacite (wt %) divided by the mean nickel content expressed as a percentage.

§ Divided by a same quantity for bulk meteorite.

TABLE 6

Inert Gas Content of the Fayetteville, Miller, and Paragould Meteorites*

	^{3}He	^{4}He	^{20}Ne	^{21}Ne	^{22}Ne	^{36}Ar	^{38}Ar	^{40}Ar	^{84}Kr	^{132}Xe
	(10^{-8} cc STP g^{-1})									
Fayetteville Dark Matrix	422	1410000	8230	40.7	698	281	60.0	6930	0.17	0.10
Fayetteville Light Clasts	62.0	1770	48.5	10.8	14.0	3.16	2.11	5980	0.555	0.022
Miller	6.03	1820	1.93	1.06	1.29	1.43	0.44	6095	---	---
Paragould	24.61	168	5.81	5.43	5.99	2.53	0.92	730	---	---

Sources: Contents obtained by averaging the data in the compilation by Ref. 8 in Table 1, and from O. K. Manuel and P. K. Kuroda, *Journal of Geophysical Research* **69**, 1413–1419.

*$\frac{1}{100}$ millionth cc of gas per g of meteorite.

TABLE 7

Compositional Data for Arkansas Iron Meteorites and for the Newport Pallasite*

	Ni	Co	P	S	C	Cu	Ga	Ge	Ir	References (See Table 1)
	mg/g			µg/g						
Cabin Creek	~80.0	---	~3.0	---	---	---	---	---	---	2
Calico Rock	54.5	4.5	2.8	50	135	140	57.3	185	8.6	5, 4
Hatfield	No data									
Joe Wright Mtn.	91.0	5.1	4.7	50	140	---	20.1	35.5	0.015	11, 5
Newport	107	5.0	---	---	---	---	17.5	31.2	0.16	35
Norfork	78.8	5.0	1.4	70	20	170	20.4	41.2	0.3	15, 5
Sandtown	80.9	---	---	---	---	---	21.0	41.4	1.4	11
Western Arkansas	72.6	---	---	---	---	---	1.97	0.100	2.8	33

*See Table 3 for explanation of concentration units.

TABLE 8

Definition of Shock Facies a–f

Facies	Olivine Texture	Plagioclase Form	Melt Pockets	40 AR	Highly Volatile Trace Elements	TL Sensitivity (Dhajala = 1)
a	Fractured	<50% deformed	None	Normal	Normal	0.8–11
b	Fractured and undulose extinction	---"---	---"---	---"---	---"---	---"---
c	---"---	>50% deformed	---"---	---"---	Bi, Ag, Zn <0.8 normal In, Tl, Cd <0.5 normal	0.07–3.0
d	Fractured and mosaic extinction	100% deformed or maskelynite	Present	~0.1 Normal	Bi, Ag, Zn <0.5 normal In, Tl, Cd <0.1 normal	0.02–1.0
e	Fractured, mosaicked, and marginal granulation	>50% maskelynite	---"---	---"---	---"---	---"---
f	Recrystallized	100% maskelynite	---"---	---"---	---"---	---"---

Sources: R. T. Dodd and E. Jarosewich (1979), *Earth Planetary Science Letters* **44**, 335–340; T. M. Walsh and M. E. Lipschutz (1982), *Geochimica Cosmochimica Acta* 46, 2491–2500; D. W. G. Sears, N. Bakhtiar, B. D. Keck, and K. S. Weeks (1984), *Geochimica Cosmochimica Acta* **48**, 2265–2272.

How to Recognize a Meteorite

Most meteorites are heavier than common terrestrial rocks. The stony meteorites are about one and a half times as heavy, while stony-iron and iron meteorites are considerably heavier. Meteorites are generally irregular in shape, but with rounded edges. Freshly fallen meteorites are black in color with textures on the surface caused by the flow of molten material during atmospheric passage. However, such freshly fallen meteorites are rare and found usually only after the fall itself has been witnessed. Usually, the meteorite has become brown due to weathering and sometimes may reveal a surface scale. Meteorites are always compact, not porous or hollow. The photographs in Figs. 1a, 4, 5, 9, and 17 give some idea of the external appearance of meteorites. A good test of authenticity is to grind a suspected meteorite on an abrasive wheel. All but a few meteorites contain grains of metal that appear as shining flecks on the abraded surface, or they are made almost entirely of metal, in which case the whole surface appears shiny. Most meteorites are magnetic to some extent.

Most major museums, including the University of Arkansas Museum, have personnel capable of recognizing a meteorite and are happy to make an identification free of charge. Send a sample to the museum, with a description of the circumstances of its find.

Further Reading

Buchwald, V. F. *Handbook of Iron Meteorites*. Berkeley: Univ. of California Press, 1975. Technical.

Dodd, R. T. *Meteorites: A Petrologic-chemical Synthesis*. Cambridge: Cambridge Univ. Press, 1981. Technical.

Hartmann, W. K. *Moons and Planets*. 2nd ed. Belmont, Calif.: Wadsworth, 1983. Chapter 6. Introductory.

Heide, F. *Meteorites*. Chicago: Univ. of Chicago Press, 1964. Popular.

Hutchison, R. *The Search for Our Beginning*. London: Oxford Univ. Press, 1983. Introductory.

Krinov, E. L. *Principles of Meteoritics*. New York: Pergamon, 1960. Technical.

———. *Giant Meteorites*. New York: Pergamon, 1966. Semi-technical.

Mason, B. *Meteorites*. New York: Wiley, 1962. Technical.

Nininger, H. H. *Out of the Sky*. New York: Dover, 1952. Popular.

———. *Arizona's Meteorite Crater*. American Meteorite Museum, Sedona, Arizona, 1956. Popular.

Sears, D. W. *The Nature and Origin of Meteorites*. New York: Oxford Univ. Press, 1979. Introductory.

Wasson, J. T. *Meteorites*. New York: Springer-Verlag, 1974. Technical.

———. *Meteorites: Their Record of Early Solar System History*. New York: Freeman, 1985. Technical.

Wood, J. R. *Meteorites and the Origin of the Planets*. New York: McGraw-Hill, 1968. Introductory.

Index